KNOW ALL ABOUT

REDUCE
REUSE
RECYCLE

KNOW ALL ABOUT

REDUCE REUSE RECYCLE

Yoofisaca Syngkon Nongpluh

Guy C Noronha

ISBN 978-81-7993-391-6
First Reprint 2015

Published by
The Energy and Resources Institute (TERI)
TERI Press
Darbari Seth Block
IHC Complex, Lodhi Road
New Delhi – 110 003
India

Tel.	2468 2100 or 4150 4900
Fax	2468 2144 or 2468 2145
	India +91 • Delhi (0)11
E-mail	teripress@teri.res.in
Website	www.teriin.org

Book design by Sonali Lal
(Sonalilal@gmail.com)

Printed in India

Dedication

"We would like to thank all those who believed in us and 'never uttered a discouraging word' as we followed the writing dream and supported us through the agonies ..."

Contents

Preface

In our initial discussions, we had decided to title the book as "The 3Rs". The title, like the book, underwent several reviews and changes in direction. As it goes to press we still refer to it in the old (now affectionate) term, but have changed the name to *Know All About: Reduce Reuse Recycle*.

Having lived in the once quiet hill towns of Bhimtal and Shillong, the subject of the book evoked strong feelings in us. Watching the population influx and tourist invasion, we could not help but notice, and be dismayed, by the fallout of this "growth and development". Besides the rapid degradation of the environment, disappearance of forests and the very noticeable loss of peace and tranquillity one of the ugliest consequences of growth and change is garbage! The vast amounts of plastic soft drink bottles, empty chips and biscuit bags lining the once picturesque and winding roads is deeply saddening. Riding up to our homes used to be a pleasure, not anymore.

So writing this was not just an exercise in research or academic fulfilment—it is deeply personal. It is also heartening that in the course of exchanges with the editors of TERI, the book slowly but happily took the direction it now does. It addresses ordinary people—without gobbledegook, technical information and charts—trying to show them how our everyday actions have huge consequences beyond our individual visions. We have tried to (we hope without being too preachy) show how change even on the micro-personal level can make a huge difference.

We hope that this book goes on to make that difference.

Acknowledgements

Patience! That has been the most outstanding virtue of everyone at TERI who has had to deal with us in the course of putting this enterprise together. The researching and writing of this book has taken almost two years!

We would like to thank Madhu Singh Sirohi, who first had the confidence to involve us in the book. She has guided us in developing the initial structure of the book and encouraged us to work on it during every spare moment we had.

To Arshi Ahmad, who pushed us to rework the first chapter (five times!), yet never allowing us to give up. She often nudged, cajoled, and regularly reprimanded us for our tardiness. Her suggestions have been invaluable.

We would like to thank Arani Sinha, once an Associate Editor at TERI Press, who had to put up with our tardy and often missed delivery schedules. It is to her that we owe a huge debt of thanks for understanding and sharing our vision of how the book should turn out.

Then there is Hemambika Varma, Editor, TERI Press, who has managed (despite our frequent disappearances) to get us to complete the writing, reviewing and editing of the book. She has been instrumental in finally putting this book to bed. Thank you!

Finally, we thank TERI for giving us this opportunity to become "authors". The organization has shown great faith in our being able to provide a work that will, hopefully, line up with many of their other great publications.

Introduction

How do you go about saving your home and prevent it from turning into a garbage heap? You clean up, collect the trash and dump it someplace else. That, unfortunately, is the start of the whole problem. Distressingly, there is no place that we can conveniently dispose off our waste without it coming back to haunt us.

The "home" we will be talking about in this book is our planet – Earth. There is no cosmic garbage landfill for our waste. What we generate stays with us. Worse still, the problem will not go away by just wrinkling our noses or holding our breath as we drive past. At the rate society, worldwide, is generating trash—garbage tips will soon be a thing of the past. Waste will eventually become a part of the scenery!

Sceptics and people with an "it's not my problem" attitude will probably think that there is exaggeration here; that things are not as bad as they are made out to be; that these are just the shrill cries of environmentalists hawking their doomsday stories. Several events in the recent past have done nothing to enhance the reputation of environmentalists and green activists.

The perfect pictures of screensavers and desktop wallpapers will rather quickly be just that – virtual pictures and the only reminders of what once was. One of the joys of a railway journey through India used to be staring at the scenery as trains made their leisurely way through hills and valleys and past quiet little rural towns. Today, the pleasure is sadly contaminated. There is a depressing sameness to the countryside now.

Every railway station and miles and miles of track are littered with castaway polythene bags, styrofoam cups, chips and biscuit covers, and soft drink bottles. These are the new components of scenic design. This does not even take into account the roadsides and the highways. Plastic and its various *avatars* are the new art materials and every one of us is the artist.

Vast swathes of oceans are just floating junkyards. Tiny innocent and achingly beautiful islands find their beaches covered with plastic, sewage, and other undesirable waste. Rubbish – not of their making – generated and dumped thousands of kilometres away land up on their shores. Just imagine neighbours from your residential colony dumping their refuse on your doorstep. That is exactly the plight of many islands in the Indian and the Pacific oceans.

What is done by people – you and I – on a small, individual scale has global repercussions. Funnily enough waste generation and its disposal or (as in some of our cities) non-disposal is one problem that can be clearly, directly linked to our everyday actions.

As children we start out by learning the 3 R's – reading, 'riting and 'rithmetic. We never stop learning, and the new lessons that are just as crucial are the new 3 R's – reduce, reuse and recycle. The actions implied by these three words will be the focus of discussion in this book.

The contents of the book are an attempt to trace the origin, growth and development of human generated waste to better understand the problem. More essentially, it contains several simple and everyday solutions to an avoidable disaster. The garbage train is racing towards us at an alarming speed but it is still not too late to slam the brakes and avoid an ecological wreck.

Life is all about change and "change is constant". To survive we need to adapt and be ready to change our thinking. If we fail to wake up and smell the stench of waste and its ally—pollution—our species and probably the planet could well be consigned to the scrap heap of history.

A load of rubbish — what a waste!

Once upon a time …

There was no rubbish. There was no garbage. And, before human beings came along, Mother Nature managed just fine. She did not waste anything. Everything was recycled and reused.

It is no surprise then that nature leads the way in recycling. She has her own effective mechanisms; and we can learn a lot from her only 'if' we are willing.

When animals die, they become food for other animals and insects and the remains decompose. The same happens to plants, which decompose and dissolve into the earth. This, in time, becomes the nourishment and energy for the next generation of plants and animals.

Two of nature's most outstanding recycling abilities can be found in air and water. These are the two most important components in the circle of life on our lovely blue planet. Without these two basic elements the earth would not be the place it is, nor would we be able to live in it. Let's see how nature continually recycles air and water so that life can continue.

Then came the Homo sapiens

This idyllic situation began to change slowly with the arrival of Homo sapiens. Humankind started to produce waste that did not immediately integrate into the environment.

In the early days, when humans were hunter-gatherers, they lived in harmony with nature. At that time man used and re-used natural objects and tools till they virtually disappeared. The waste they produced was largely ash from fires, bones, bodies, plant, and vegetable waste. All of these easily dissolved into the earth, enriching the soil in the process. The waste that did not go back to the earth was small in quantity. More importantly, it was natural, like tools and hunting weapons made from stone. Early humans repaired and reused nearly everything.

Archaeological excavations of sites that date back to the pre-settlement times in human history show that the waste generated then contained only small amounts of ash, broken stone tools and, at times, pottery. Moreover, the human population was very small then.

One of the main reasons why we do not find much evidence of waste or garbage of this period was the low density of the human population. Humans lived and moved in small groups, often far from one another.

This state of affairs did not last long. Man soon gave up his wandering lifestyle and began to live together in settled communities, then villages, which grew to become towns and later cities. The fallout of this "community living" was the accumulation of waste at one place. Soon enough, the increase and build-up of rubbish became a problem.

It is this reason why ancient cities were not very pleasant places to live in, despite the security they offered to its citizens. The idea of civic consciousness only came thousands of years later. At the time of the great civilizations of Ancient Rome, Greece, Troy, and China, their cities had begun to have serious garbage problems.

The domestication of mankind took centuries and the changes in his living pattern triggered the waste problem. As the population increased and the size of cities grew, so did the concerns about mounting waste. Alongside grew the pressing need for proper waste disposal.

In order to avoid confusion, we need to find out exactly what is waste, rubbish or garbage.

The nature of waste: What is it?

Like most things in this universe, even the nature of waste has changed over the centuries. Let us find out what waste or garbage contains today.

What exactly is rubbish?

- Rubbish is *everything* that you throw away or no longer have a use for.
- Rubbish is anything from an empty biscuit packet to a broken toy.
- Rubbish can be solid, liquid or gas.
- Often people use the word "waste" or "garbage" when talking about rubbish.

There are three different sorts of rubbish or waste:
- Domestic rubbish/waste - from households.
- Industrial and commercial rubbish/waste - from factories, offices, shops, and schools.
- Hazardous rubbish/waste-mostly produced by small factories and even large industries. This needs to be disposed off very cautiously in order to prevent pollution. A good example is chemicals used in the making of paint.

Another emerging category of waste is nuclear waste, which is even more dangerous and requires scientific disposal. E-waste is also a potential hazard that has many dangerous effects.

Waste is composed of different substances and materials.

- Biodegradable waste - it breaks down naturally, integrates back into the environment, and eventually disappears. For example, leftover food and paper.
- Non-biodegradable waste - it does not break down naturally into harmless ingredients. Some examples are soft drink cans and plastic bottles.

What we use, and how it is produced, has a direct relation to the amount of waste we generate. How we dispose of it is a serious problem today; and so is the impact of waste on the areas around where we live. The disposal of garbage has a long and dirty history indeed.

History of waste disposal

The rapid increase of population in cities and towns has led to a struggle in dealing with the problem of waste disposal. The history of human civilization is littered with attempts to create ways to deal with garbage and its safe disposal.

The first evidence of a landfill site goes back more than 5,000 years ago. The ancient and fabled bronze-age city of Knossos, the capital of Crete, had large pits where waste was dumped. The garbage was then regularly covered with earth.

China gave us the compost heap. It was a regular and widespread practice by the Chinese to turn their rubbish into compost. They were doing it more than 2,000 years ago. Europe, around the same time, had scrap recovery systems in place.

The city that gave the world democracy—Athens—was also grappling with the problems of waste disposal. Over 2,500 years ago, administrators of the city created a landfill site a mile beyond its gates, and ordered that the city garbage be dumped only there.

The large amounts of stinking waste accumulating in the towns and cities around Britain led to the passing of a law in AD 1297, demanding all householders keep the front of their houses clean and free of rubbish. However, there were many who disregarded the law.

Many more instances in our recorded history show how the creation of waste and its disposal has caused problems.

You and I are responsible

The poor state of waste management and disposal is a worldwide phenomenon. Yet the causes lie in the little things that we do or don't do every day. "What harm could come from that empty bag that contained potato chips I flung out the train window?"

"How will that can of soda I threw out from the car on my trip to the lovely green Himalayas mess up the place? I only do it occasionally."

It makes a difference.

We mistakenly think that our "once in a while" litter does not matter. Cumulatively every single action of ours leaves an indelible mark on the planet. Yet, we like to think we are responsible individuals, and even convince ourselves that we are conscientious.

Every day when we drive to work or drop off the kids at school, we see evidence of our wastefulness. We can see exactly what we are doing to our neighbourhoods and towns. It is an all too common sight to find plastic bags and bottles littering our roadsides.

It takes 1,000 years for polyethylene bags to break down; plastic bottles take 450 years and most plastic bags can take as much as 20 to 100 years to degrade.

Soft-drink aluminium cans take 200 years to degenerate.

It takes 50 years for styrofoam to dissipate.

We are blind to the huge problem that each of us creates. Yet we continue adding to it every day. We bag it; the house cleaner takes it away and dumps it in the local dustbin. We, in our heads, have been good and responsible citizens. However, we forget that what we dispose of does not disappear into the air.

Far from it!

Our garbage just moves to another place. Sometimes quickly and efficiently, other times not for weeks or months. In our shortsightedness, we just do not realize how long it takes before garbage ceases to exist.

Another example is a glass bottle that we throw away. Left to itself, the bottle takes ONE MILLION years to perish! We all know that ordinary sand is used to make a glass bottle. Yet the life of a bottle is longer than it has taken humans to evolve as a species.

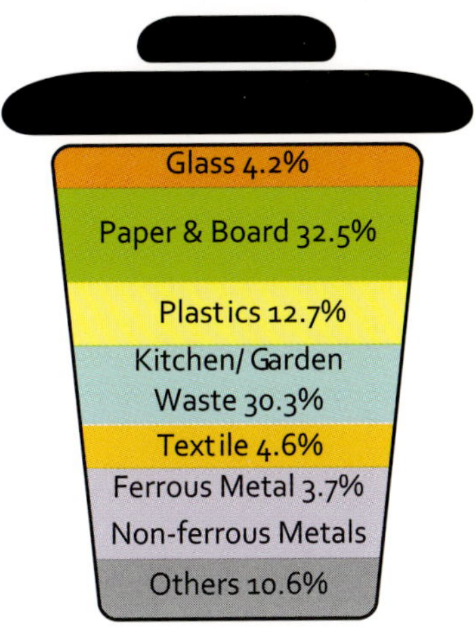

We also have the rather naïve belief that it is only factories and power plants that are the sole cause of waste creation. It is actually both industry and us—consumers, buyers, users—who are equally responsible.

To put things in perspective, imagine you had a house with a nice backyard. The average Indian household generates disposable waste of about one kilogram a day. That means every year you and your family produce about 365 kilograms of garbage, the average weight of five adults. Now imagine your pretty backyard piled with rubbish. Pretty soon, it is going to be a mountain of trash.

Athens

The Athenians introduced democracy to the world. The city dwellers took their "people power" very seriously but they dumped their rubbish onto the street without a thought about its impact. Things got so bad that around 500 BC, the Athenian city fathers were forced to issue a law that banned people from throwing their garbage into the city streets.

Troy

The fabled city of Troy was destroyed by the Greeks. Troy could very well have been ruined by its own rubbish.

In the city of Troy, people would simply dump their garbage outside their own doorstep! There was no garbage management system then. When the smell of the garbage became unbearable, the people would cover it up with mud.

This build-up of garbage, and the covering of mud, forced people to raise the roofs and doorways of their homes.

The picture is even more staggering in developed countries. For instance, every year, people in England and Wales throw away about 8,000 tonnes of garbage. Do you see just how elephantine the waste problem is?

The waste problem is not static. It is growing every year at the rate of about 1.4-2 per cent. There are many reasons for the fast growth of waste. One of the major factors is the economic growth of many developing nations driven by the improvement of their living

standards; and a rapidly growing population. Ironically, as we achieve better living standards, we create more garbage. Weird but true!

It may come to pass; all that survives of our civilization and culture is our garbage. This is definitely not what we want to be remembered by.

How waste became such a big deal

Human population has grown rapidly through the ages. There are twice as many people as there were less than 100 years ago. The phenomenal population growth has, in turn, caused a dramatic increase in waste. This has given rise to other problems, mainly health and pollution related.

Have you ever heard of basic human necessities? Strange how yesterday's luxuries become today's necessities. Our constantly growing "needs" begs the question, "When is enough, enough?" The answer must be determined now!

Industries have grown to satisfy our hunger and demand for newer devices, tools, and machines. The Industrial Revolution came about to produce more things for ever more people. As a result, more waste was generated by factories and more rubbish produced by people.

In just a little over 100 years the things we consume and produce has changed very dramatically. A brief timeline will give you an idea of just how our waste has changed over time and as a direct consequence, how its disposal causes ever more problems.

Impact of the Industrial Revolution

The late 19th century, Industrial Revolution saw factories produce more and more to meet the rising demands of people. This led to the increase in waste both industrial and domestic. What was revolutionary about this period was the way manufacturing changed. It was the time when the processes changed, volumes increased dramatically and factories were "invented"'.

Along with the revolution in the way we produced goods came a new and sinister problem—industrial waste! This added to the already existing problem of household waste.

The 19th and 20th centuries saw landscapes and cityscapes change dramatically. Skyscrapers started to dominate the skyline, while household garbage and industrial waste took over the spaces in between. Cities in the US and Europe were increasingly clogged with industrial waste. The filthy waste also led to the spread of diseases and pollution of the air and water.

The situation created by the Industrial Revolution was not really recognized or understood until the middle and late 20th century. By then the impact of environmental damage caused by garbage of all kinds and from all sources began to be felt.

Another product of the Industrial Revolution was the invention of the tin can—the father of packaged goods. It was also the first of packaging materials that would start to fill garbage piles and increase disposal problems.

The aftershock of the Industrial Revolution, in the form of waste, was enormous and continues till today.

The emergence of plastic

Around the 1950s came a most horrifying development—the increased varieties and applications of plastics. No one foresaw or understood the dangerous non-degradable make-up of plastic and its toxic or poisonous side effects. The ever-rising consumption of plastic and other kinds of waste has brought us to the stage where our planet is looking increasingly like a chain of garbage heaps.

Plastic: it's emergence, usage, and danger Plastics and their usage are not new, relatively speaking. Polystyrene, a plastic polymer, was discovered in 1893 by the German pharmacist, Eduard Simon. However, it was not until 1930, a whole hundred years later, when a German company started to commercially manufacture polystyrene.

Plastics in all its forms, shapes, sizes, and colours are everywhere. It is in every product or merchandise that we use. From packaging to electronics, this synthetic creation of man is a marvel. Such is its versatility that today we cannot imagine life without plastic.

Plastic is also probably the greatest threat to the health of our planet and our own well-being.

The bigger picture

The bigger picture is often lost to us. Rather we are unable to see it because it is so huge. Rubbish, waste, garbage, call it what you want, damages the environment in different ways and at several levels.

The link between waste (disposal) and health—of the earth and ours—is clear and visibly noticeable. The harm caused by waste—industrial or domestic—is pollution and environmental degradation.

Pollution is caused by several factors and it affects the water, soil, and air. The damage is long-term and the effects can spread to areas over great distances. It also takes a long time to clean up.

An example close to home is the waste that Delhi dumps into the Yamuna. The pollution of the river as it flows past the capital affects everyone downstream. Fish die, the water is not potable. Even farmers, hundreds of miles away, cannot use it for irrigation. Birds that feed on the fish or by the riverbanks die. The waste has entered our food chain as well.

Ancient Egypt

It is a fact that the fabulous ancient Egyptian civilization flourished as it did because it was supported by the River Nile. This beautiful river nourished the land along its banks as it flowed along.

The treasures of the Egyptians were nothing compared to the importance of the Nile and yet, the Nile was also very convenient for their garbage problem. They simply dumped their waste into it. They treated the Nile as a convenient, watery trash-disposal system.

The big picture – from the ocean

The United Nations Environment Programme (UNEP) in its 2009 report on marine litter stated, "…. it is an economic, human health, and aesthetic problem. It poses a complex and multi-dimensional challenge with significant implications for the marine and coastal environment and human activities all over the world".

Choking To Death

Around 100,000 whales, seals, turtles, and other marine animals are killed worldwide after swallowing plastic, according to a July 2004 report by Planet Ark, an international environmental group.

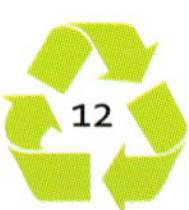

Plastic Bags

Environmentalists estimate that between 500 billion and one trillion of these bags are used every year worldwide.

The report holds "various human activities, an inadequate understanding on the part of the public of the potential consequences of their actions" mainly responsible for the international, regional, and national problem. It also specifically identifies cigarette smokers as being huge contributors to marine garbage.

> ## It wasn't the Exxon Valdez captain's driving that caused the Alaskan oil spill. It was yours.
>
> ~Greenpeace advertisement, *New York Times*, 25 February 1990

Another gigantic component of marine garbage are plastic bags and bottles. It makes up the largest part of sea litter around the world. The UNEP report found that "every year, marine litter results in tremendous economic costs and loses to individuals and communities around the world. It can spoil, foul, and destroy the beauty of the ocean and the coastal zone".

Most of the litter in the oceans comes from garbage dumps close to or on coasts and riverbanks. Other big contributors are merchant ships and fishing boats that illegally dump their trash at sea.

Achim Steiner, UNEP executive director, says, "Some of the litter, like thin film, single-use plastic bags, which choke marine life, should be banned or phased-out rapidly everywhere (because) there is simply zero justification for manufacturing them anymore, anywhere."

The marine garbage problem is vast and terribly unfair. For instance, Seychelles has a huge garbage problem during the monsoons. The winds and wave action drive garbage from other countries to their shores. This beautiful nation of islands produces hardly any garbage of its own. So garbage from Western Australia makes its way across the vast Indian Ocean to the beautiful coastline of eastern parts of South Africa.

The most famous rubbish tip in the world is in the Pacific Ocean. Some estimates of the patch report it to be roughly the size of Texas, which, at 696 200 sq. km area, is the second largest state in the US. Just picture an enormous expanse of ocean covered with plastic trash ...!

The waste dumped by factories into rivers, rubbish from homes carted off to landfills, affect human health and the surrounding areas in several ways. The poor storage or disposal practices cause the toxins from rubbish to seep into the earth and degrade it, which presents an even greater threat because it then becomes impossible to fix.

Here, in our own backyard, South Asia, the growing shipbreaking industry has become a major source of toxic marine waste.

We must realize that garbage is a big deal. Waste could engulf us.

Surface water pollution

Waste of all kinds causes changes in the water chemistry. Polluted surface water (rivers, lakes, and ponds) affects the surrounding areas and poisons the environment and ecosystem. It has an impact on the health of all creatures across the food chain. In other words, birds, animals, and humans are affected when they feed, drink, and bathe with polluted water.

Tiny organisms in the water that support plants and fish accumulate and concentrate the poisons, which are transferred to the fish, animals, and humans that feed on them. The dosage of poisons from the infected creatures is much higher than what you consume from direct ingestion.

Water

Water is the essence of all life on our planet—no exceptions. All living creatures need water to stay alive. Some like fish, crabs, whales and other aquatic animals even need to live in water to stay alive. And nature, quietly and continually keeps recycling water. This action is called the Hydrological Cycle. One could also call it the water circulation and recycling system.

This system is powered by the sun. The heat produced by the sun is the main force behind this process.

Rainfall is the most obvious and visible part of this water recycling system. Water from lakes, rivers, seas and the oceans is heated by the sun. It evaporates and partially condenses to form clouds.

The clouds, in due course, releases this moisture in the form of rain. The cycle starts, all over, again. When the rains fall it nourishes the plants and trees. It fills the rivers and lakes that give humans and all living creatures the water they need.

Plastic does not quickly decompose, and requires high energy ultraviolet light to break down. This process is known as photodegradation. When plastic photodegrades it turns into small micro-particles, which enter the river networks, into the water supply, and is then consumed by people.

Groundwater pollution

Chemicals from garbage and other industrial wastes soak into the earth and contaminate the aquifers and underground lakes. These underground water resources then feed into surface lakes and rivers. In

Soil pollution, also commonly known as soil contamination, is a condition that occurs when soil loses its structure, biological properties, and chemical properties due to the use of various man-made chemicals and other natural changes in the soil environment. This form of pollution is more common in developed countries.

Air

All green living things like trees, grass and plants breathe. When they "inhale" or breathe in they take in carbon dioxide and sunlight through their leaves. Along with minerals from the soil, the trees and plants 'cook' the carbon dioxide using sunlight. This is their food. This process of making food is called photosynthesis and goes on all the time. And when the trees and plants breathe out they give out a life-giving gas—oxygen!

This is a really smart exchange. Trees utilize and use what is harmful to most living creatures – carbon dioxide. In return, they give us oxygen. This is just one example of how nature brilliantly recycles. In doing so, she ensures that everyone benefits.

A simple and strong reason why we need to keep the trees and plants. If we don't – we could all very well run out of oxygen!

many regions of the world, groundwater is extracted and used for drinking, bathing, and agriculture.

Soil pollution

When toxins from waste get into the soil, they harm, and in many cases, kill plants and trees. This happens when plant roots take in the poisons. Coming into contact or inhaling the air around contaminated soil causes health problems for animals and humans.

Air pollution

Household and industrial waste—chemicals released into the air causes air pollution. These airborne chemicals cause breathing problems when inhaled. Some chemicals cause skin problems. They also affect plants when they ingest the polluted air.

Education and caring

There is nothing like a little caring and focussed education. These two tools can help us to do something (however small) to clean up our world. Every little bit helps and we need to protect our environment now!

We have to understand that we are responsible for our actions. Part of that responsibility is to use less. Think carefully before using and, even more carefully, while disposing of.

> **Thank God men cannot fly, and lay waste the sky as well as the earth.**

> ~Henry David Thoreau

The above quote has to be the most ironical statement ever!

Just got to have that!

We humans share a very quirky trait with the magpie. When we see a shiny new object we want it, and badly!

This hunger for more and newer things is one of the biggest reasons for the pile up of waste and garbage all around the world. Goods and products are made so as to become either useless or outdated very quickly. Very often, it is more economical to buy a new product rather than repair an old one.

What is even more worrying is that the global economic system is based on ever more consumption. More consumption means more waste, more pollution. It is high time to protect the environment and preserve the earth's resources. That's why deliberately making goods that have a

very short lifespan is not sustainable. It exhausts resources and creates more pollution in the process.

The concept of manufacturing products that last for a very short time is known as "planned obsolescence". The following examples will give you an idea of how the concept of planned obsolescence is creating waste. This is a good and profitable strategy for manufacturers. And we buy into it. In the process we are "messing" with the future of our planet.

Smartphones... Really!

A very good example of the "buy new" approach is that of a recent model of a leading smartphone manufacturer. Greenpeace carefully examined the phone and found that its design made it very difficult to take out the battery and replace it. You are compelled to go out and buy a new phone.

Fashion—Classic Case of Use and Discard

The garment industry is the best example of built-in obsolescence. Fashion by definition, means that last year's dresses, for example, are designed to be replaced by this year's new models. In many cases the dress worn at the last party is obsolete!

Computed to be Useless—Quickly

Planned obsolescence is how the computer industry survives. New software is designed to be incompatible with previous versions. Older programmes do not work when using new software; in other words, the new versions can read all the files of the old versions, but not the other way round.

Intelligent Semi-conductors

The strategy of making products quickly obsolete is the hallmark of the semi-conductor industry. A leading producer of computer chips was working on the next generation of chips even before they had started selling the last one. This has led to a situation where computers, gadgets, and other devices are obsolete even before they are in the market!

Everyone is a junkie

The seemingly endless and massive volumes of waste produced across the world are astounding. It is not just big industry that is the cause. It is also ordinary people like you and I. The amount of waste generated is often closely tied up with the income levels and lifestyles of people. For instance the average US family is reported to throw away about a tonne of garbage a year. The Indian family, while not throwing away as much, is not too far off. We, Indians, generate almost half a tonne of rubbish a year, per family. Most of this comes from the richer urban Indians. The poor generate less waste.

As India's economic growth rapidly increases so does its waste. The management and disposal of waste is assuming alarming proportions, as also in China. Reliable and complete data on garbage generation in India is difficult to get. This is especially true of most developing countries. Also, definitions of garbage/waste differ from country to country, which makes battling the menace extremely difficult.

Every day, and in every way, we use newer products dangerous to the environment. The vast majority of our daily purchases are either packaged or made from plastic. It is bad for the environment in two ways. Plastic takes hundreds of years to break down and it is made from petroleum (itself a rapidly declining resource) by-products. Remember this the next time you take a plastic bag or throw away one.

Some harmful garbage

In our modern enlightened state, we all want to live healthy. So we pick up a "fresh and healthy" tetra pak of orange juice that is now available at every corner store and supermarket.

However, right from start, that attractive looking pack of orange juice has damaged the environment. The entire production process from start to finish uses more than 900 litres of water and two litres of petrol, for the irrigation, machinery, and trucks. Many pesticides are used to

produce commercial oranges, which work their way into the drink and into our bodies.

A fresh orange bought locally and squeezed at home is far less damaging. You could then put the orange skins into a compost pit and refresh the soil. This is far healthier for you and the earth. Another example is commercially produced cereals. It takes two litres of petrol to produce a kilo-box of cereal!

"Food Miles" is a term used to describe the distance your food travels from where it is made to where you eat it. According to the US National Sustainable Agriculture Information Service, this distance has been increasing over the last fifty years. Processed food now travels an average of 1,300 miles!

Transporting food is one of the fastest-growing sources of greenhouse gas emissions, according to the *World Watch Institute*. Each year, 817 million tonnes of food is shipped around the planet. The result is that a basic diet of imported products can use four times the energy and produce four times the emissions of an equivalent domestic diet!

Food miles affect the environment because they contribute to greenhouse gases emissions from the transporting vehicles, most notably carbon dioxide. Greenhouse gases are a major contributor to climate change.

To make our homes spick and span and our clothes spotless, we use ever more powerful cleaning products. The manufacturers assure us that these detergents can remove the most stubborn stains.

The chemicals may clean the floors and clothes but they can also harm you. The detergents are made from chlorine and phosphates. When you flush them down the drain, they make their way into waterways and the soil. Cleaning  products are poisonous to plants and wildlife. They also pollute our rivers and lakes, killing fish and endangering marine life.

Cleaners or Killers!

Dish cleaners, clothes stain removers, metal, furniture, and floor cleaning chemicals, carpet cleaners, glass and window cleaners and oven cleaners are some of the products that fall under the category of all-purpose cleaners. The ingredients of all-purpose cleaners include ammonia-based chemicals, bleaches, sodium hypochlorite, and trichloroethane solvents.

Short-term exposure to sodium hypochlorite, found in chlorine bleach, if mixed with ammonia, releases toxic chloramine gas. This gas may cause mild asthmatic symptoms or more serious respiratory problems. Ammonia, found in glass and other cleaners, causes eye irritation, headaches, and irritation in the lungs.

The chemical components in phenol and cresol—disinfectants—can cause diarrhoea, fainting, dizziness, and kidney and liver damage. Chemicals in spot and carpet cleaners can cause liver and kidney damage and induce cancers.

These are just a few examples to show that new technologies and new products are creating waste materials that are harmful and dangerous

both in the short and long run. We are breeding a situation where there are serious risks to our health, the environment, and to all life forms on the planet.

> **The packaging for a "microwave" dinner is programmed for a shelf life of maybe six months, a cook time of two minutes and landfill dead time of centuries.**
>
> ~David Wann, *Buzzworm*, November 1990

New is not necessarily better—despite the claims on the product boxes!. The products are either difficult to recycle or are poisonous. More and more items of everyday life do not decay and harmlessly dissolve back into its components. They end up in the garbage dump.

We need to clean up our act

Our lifestyle involves getting newer, updated, upgraded, improved or enhanced products. It is a very important reason for the deterioration of our environment. The problem is particularly bad in the developed and industrialised countries. It is also assuming Himalayan proportions in emerging economies such as India and China.

A grim fallout of this has an adverse impact on the climate. It is also causing friction between the industrialised nations and those that are seeking to develop their own economies. Every country is looking to produce and consume more in order to "improve" the living standards of their people. But is all this really "improving" the standards?

We need to change our patterns of consumption and also reduce or stop using products that produce dangerous waste. We need to stop being wasteful in our production processes.

We must reduce the use of energy and materials that end up being garbage. Better still learn to say "No".

Wealth and waste—a close connection

In India, people in the high income bracket (₹ 8,000 per month and above) generate about 800 gm of waste per capita every day, while the low income bracket (₹ 2,000 or lower) generate only 200 gm of waste per day.

[Srishti and The Energy and Resources Institute (TERI)]

Waste around the World

GRID-Arendal is an official United Nations Environment Programme (UNEP) collaborating centre, supporting informed decision-making. This is what its report on waste generation, Vital Waste Graphics, says:

"The Basel Convention has estimated the amount of hazardous and other waste generated for the years 2000 and 2001 at 318 and 338 million tonnes respectively. However, these figures are based on reports from only a third of the countries that are currently members of the Convention (approximately 45 out of 162). Compare this with the almost four billion tonnes estimated by the Organisation for Economic Co-operation and Development as generated by their 25 member countries in 2001 Environmental Outlook, OECD."

Even though the reporting is uneven and data, in most cases, unavailable it still makes for a horrid picture.

Why we have to say "NO"

Here is a really scary picture. All over the world, two million plastic bottles are thrown away every five minutes! That is one huge garbage problem. It also means that those bottles represent over 3,375,000,000 litres of crude oil that went into their manufacturing.

Drive from home or to the office or take a trip to the hills and all you see are thousands and thousands of plastic bags lining the roads. Once green fields are now littered with discarded soft drink bottles, empty chips, and plastic bags. Instead of rows of gently waving crops, we have a countryside littered with plastic.

Do you remember the times you went shopping with your mother? She would take a big shopping bag and buy fish and eggs wrapped in a piece of paper. The veggies would be weighed out by the shopkeeper and put into mum's bag. When you got home all the produce would be put away but nothing would go into the garbage bin.

That is not the case today. Now meat and fish come in a vacuum-packed plastic wrapper inside a Styrofoam container. Bread and biscuits come in plastic wrappers. Most products are wrapped in boxes and packaging materials that are toxic to the environment and are not degradable.

Did you know that about 18 per cent of the money you spend on a product is actually the packaging cost? Most of the stuff we buy is not recyclable and winds its way into garbage piles that pollute our planet. Therefore, buy smart and save!

"NO" is not negative

We cannot go on buying new products and enjoying our current "throwaway" lifestyle. It is really important that we start to increase what we reuse and recycle and reduce the amount of waste we produce.

Throwaway facts

- The average American office worker uses about 500 disposable cups every year.
- Every year, Americans throw away enough paper and plastic cups, forks, and spoons to circle the equator 300 times.
- About 0.1 million tonnes of municipal solid waste is generated in India every day. That is approximately 36.5 million tonnes annually.

A simple guide on reducing

- Do not to buy heavily packed goods and wherever it is possible buy loose, unpacked alternatives
- Take your lunch to school or office in a reusable container/tiffin box
- Carry reusable carrier bags when shopping or going to the bazaar
- Reuse scrap paper and envelopes
- Buy rechargeable items instead of disposable ones, e.g., batteries and camera
- Buy products in refillable containers, e.g., Washing powders
- Buy concentrated products that use less packaging
- Look for long lasting (and energy-efficient) appliances when buying new electrical items
- Use low energy bulbs which last longer and use less energy

Now you may think that you cannot do much to change or repaint the garbage scene. That is not true. Every one of us can make a difference. The first step to making a difference is saying "no".

Saying "no" is not a bad thing and neither impossible, especially when it comes to stuff that will be trashed in the end. It does not need a huge shift in our thinking. It will not cause too much pain. All you have to do is make a small change in the way you think. You can change your life, make it better.

Saying "no" only requires that you plan just a little ahead of your next shopping trip. That's all.

> **And Man created the plastic bag and the tin and aluminium can and the cellophane wrapper and the paper plate, and this was good because Man could then take his automobile and buy all his food in one place and He could save that which was good to eat in the refrigerator and throw away that which had no further use. And soon the earth was covered with plastic bags and aluminium cans and paper plates and disposable bottles and there was nowhere to sit down or walk, and Man shook his head and cried: "Look at this God-awful mess.**
>
> ~Art Buchwald, 1970

3

Reduce — A DIET to save the world

Over consuming — Reduce

None of us like to be overweight. So what do we do? We cut down on fatty foods. We watch what we eat. We are careful with what we consume. We exercise. We go on a diet.

Unfortunately, the situation is not the same with garbage—the opposite in fact. We use too much. We generate too much waste. We end up throwing away too much.

We need to reduce. We ought to change our consumption habits. We have to watch that *"waste-line"*! We must go on a diet, this time, to save the planet. Reducing is the easiest of the "Three R's" to follow.

Just like all diets, *reducing* can be quite difficult in the beginning. You will feel that it is too hard; that your efforts are having no effect. Besides, there is too much temptation out there, which can break the best of good intentions and practices.

Pressure to package

In this era of intense competition between manufacturers, the role of packaging in marketing has expanded dramatically. The need to attract and retain customers has underscored the massive importance of product packaging. Packaging is important for manufacturers if they want to grab eyeballs. It is now de *rigueur* to over-pack.

This pressure has directly led to the increase in the use of metals, plastics, and other non-degradable packaging materials. Much of the packing is non-essential and non-recyclable.

A shiny cellophane wrap envelopes the attractively designed box that has tempting and brightly coloured pictures of the product inside. Inside the box, you very often find corrugated stiffeners that ensure the product does not get squashed or broken. There is also additional packing intended to keep the item fresh. This is overkill, in all its packaging glory!

For every tonne of garbage we generate, 10 tonnes of resources and material are used in its manufacture.

Think before you buy

So before you go out shopping take a minute to consider what you are going to buy. Remember your everyday decisions will ensure you become a vital contributor to turning our world into a prettier, better, and cleaner place.

We need to come to terms with conflicting priorities that we face, such as convenience, cost, and responsibility.

By resolving them judiciously you could become a role model for your neighbourhood and community.

A major part of the solution to the rapidly growing garbage problem lies in our own hands. Reducing should be a lot easier because it takes very little effort to reduce the amount of rubbish we produce.

> **Most of us are familiar with recycle and reusing, but how often do we think of the third R, that is, - REDUCE? "Reduce" is probably the most important of the 3R's because, if we reduced, it would limit the need to recycle and reuse.**
>
> ~Catherine Pulsifer, from The "Reduce" of Recycle and Reuse

Whether we like it or not, whether or not we can afford it, the environment is forcing us to change our habits. We have to downgrade from fuel-guzzling to fuel-efficient cars. We have to choose to buy locally grown vegetables and fruits. Let us also change in other ways.

Changing your thinking is the best way to start. Think about ways you can reduce your waste when you shop, work, and play. There are plenty of ways for you to reduce waste, save yourself some time and money. While you are doing that you are also being good to the earth.

Refusing lessens waste

Before you reduce, you could actually refuse. That makes reducing so much easier.

Here are some tips on how you can make a difference by saying "no".

Start saying "no" when the shopkeeper puts your purchase in a plastic bag. Place the items you have bought into the bag.

An even better idea is to get the old-fashioned and colourful *jhola* made from jute. Besides, you will be putting money into the pockets of the jute workers. The jute industry is also a sustainable one.

Say "no" to that new mobile phone. There is a new one out every month; you will always be behind anyway.

Refuse a receipt or invoice for your purchases **if it is not necessary**. On most occasions, we just crumple the receipts and throw it away. Save paper.

When you visit a takeaway restaurant, refuse the carry bag in which your order is placed. You only need the box that contains your food. You are going to throw away that bag, anyway.

Ask your credit card company and bank to stop sending you promotional literature. You are probably swamped with brochures and pamphlets that drop through your mailbox every month. Besides being a nuisance, no one really reads them. You could ask the bank or the credit card company to send information to you via the internet.

Refuse disposable products such as razors. Instead opt to buy much more durable and longer lasting razors. Disposable razors are the best example of how much unnecessary packing goes into a product. There is also the economic factor involved in it. Disposable razors come for almost three times the price of the normal variety.

Refuse those disposable nappies. By doing just this you will be reducing the huge amount of material that forms the garbage mountains that blot our city landscapes.

Do not buy paper napkins. You only use them once and then throw them away. Picture all those trees you will be saving by just avoiding to use paper napkins. Use cloth napkins instead. These are larger, absorb better, and make your table look nicer too.

Choose shops that sell vegetables and fresh produce items loose. The fruits and vegetables sold at shops are fresh. The added incentive to you should be that these are not packaged. It is a win-win situation. You get fresh products for less money. You also save on the packaging thereby reducing the amount of garbage you generate.

Instead of buying new toys for your children every other day, treat your children to different toys by borrowing or exchanging toys with other parents. This might be a little hard to do and your children may not be too thrilled, but worldwide experience has shown that the kids benefit from this trend. The sense of sharing is inculcated in them without much difficulty.

Do not buy those trendy seasonal clothes that last for only a year or so. Buy higher quality clothing that last longer. They save you money in the longer run.

If you put your mind to it, you will find many more things to refuse.

When you refuse, you are making a conscious contribution to global waste management. So take a stand right now. Say "NO" and help save your planet.

Diet tips: a guide on how to reduce

What can one do to reduce the amount of waste we produce every day? This is an important question because by reducing we can conserve the forests, oil, gas, minerals, and water that go waste.

Usually when buying an item, we have three things in mind.
- Requirement
- Options
- Affordability

We have to add another consideration to that list. We should ask ourselves whether our purchase would become a pollutant once we are through with it.

The reality is daunting. Consider this fact: Two million plastic bottles are thrown away every five minutes! This is happening all over the world. The bottles are manufactured from 750 million gallons of crude oil per year.

Another worrying fact is that less than 20 per cent of the disposable plastic bottles are recycled. So imagine the difference you would make if you lessened your dependence on plastic bottles, packages, and other non-degradable products.

There are many ways to reduce rubbish in the home, at school, and at work. You probably know how to cut down on the use of plastic and other polluting materials. Here are some common everyday things that you can do.

In the first instance, minimize the amount of stuff that goes into your garbage bin. This is probably the most effective of the 3R's.

Did you know?

If each of the UK's 10 million office workers used one fewer staple a day, it could save 120 tonnes of steel a year.

[Source: www.greenecoservices.com]

When you shop, buy things that will last, things that are durable, well made, and useful. If one religiously starts practising this the marketing term "consumer durables" could just start living up to its true meaning and in the best possible sense.

One way of reducing waste generation is to refuse to take unnecessary carry bags given out by stores and shops. It should become your first step when you go shopping.

Try to refuse packaging altogether. There is too much of it anyway.

Buy in bulk or choose large containers. That way you reduce a lot of packaging material. Remember that nearly 18 per cent of a product's cost is due to packaging. Therefore, you also save money.

Do not chase the latest fashion in clothes. Trends are cyclical and we all know that those previous styles keep coming back.

Avoid buying goods made from materials whose extraction or processing are especially destructive to the environment and waste intensive. A good example of this is furniture made from tropical woods. Gold jewellery is another product that is the result of an "environment-unfriendly" process.

Buy rechargeable batteries. With everything going electronic, we use batteries for a whole host of devices—from cameras to phones, to remote controls. They are used to power CD players and computers, watches, and phones as well. There is an overdose of batteries in our lives. Batteries make up a large part of our rubbish and are very toxic for the environment too. Always opt to use rechargeable batteries.

Shun overly packaged goods. The packaging is a total throw-away and just adds to the waste and disposal problems.

Where possible buy loose alternatives. Whether it is pulses, grains, fruits or vegetables buying these items in shops that sell them loose is a better alternative. Where is the need for bananas or apples to be packed in a fancy plastic net bag when the first thing you do is throw away the bag while storing in the refrigerator.

Cancel the delivery of unwanted newspapers or unnecessary magazines. Subscriptions to several newspapers may give you the idea that you are well informed and accessing different viewpoints. Valid as those arguments may be but you should re-evaluate them. Think of the trees you will be rescuing from the axe. Make a conscious effort to buy just one newspaper and choose to read the e-versions of the other newspapers.

Reduce your purchases by starting your own vegetable garden or buying locally produced vegetables. Besides the obvious advantages of ensuring that your veggies are pesticide-free and the savings you are making, there are "bigger picture" reasons. Savings come in the form of lesser transportation, which in turn leads to reduction in the carbon emission volumes.

Make some lifestyle changes; for example, lessen your reliance on fast and convenience food. This industry is highly dependent on the use of non-degradable packaging that fill up our garbage dumps.

Make a conscious effort to not buy one-time-use disposable products like paper napkins. Use cloth napkins, rags and mops in place of disposable items.

Reduce the amount of paper you use and throw away in office. Print documents only when necessary. Save copies (e-versions) of the documents instead of wasting paper in filing.

Take a homemade lunch to work instead of buying a packaged sandwich or meal. It reduces waste and saves you money. Moreover, nothing is better than home food!

- Jute is a natural fibre that is partly wood and partly textile. It is known as lingo-cellulosic
- Jute is a biodegradable fibre and has been a low cost packaging material for centuries
- Jute is notable for its silky lustre, high tensile strength, low elasticity and substantial resistance to fire and heat
- Jute is a good insulation against sound and heat and is used in the construction of houses
- Jute bags and ropes can be used again and again
- Jute is easily recycled

Impact of transportation on garbage generation

Transporting goods over long distances means that the items have to be protected against damage by handling, movement and external causes, such as heat and dust and, in the case of food, deterioration. These are some of the reasons that make companies add ever more layers of non-biodegradable packaging around their products.

Transportation of goods over great distances require production of boxes, metal cans, printed paper, labels, plastic containers, cellophane, glass jars, and a wide variety of sealing compounds. Each item adds to the size of landfills and the burden of disposal.

Reduce purchases that are packed in non-biodegradable materials such as drink cans and plastic bottles. Buy green.

Resist the urge for impulse purchases. With all the temptations beckoning from hoardings and shop windows, try to buy only what you need. That will be the acid test for your willpower or should we say "no power". Stick to your decision because you are doing your bit to reduce the height of that garbage mountain.

~Guy Debord

Once you start consciously avoiding disposable items, you will come up with some of your own ways and means to reduce waste. Your individual values and the choices you are willing to make to reduce waste, will also influence and change people around you. Become a "Reduce Leader"!

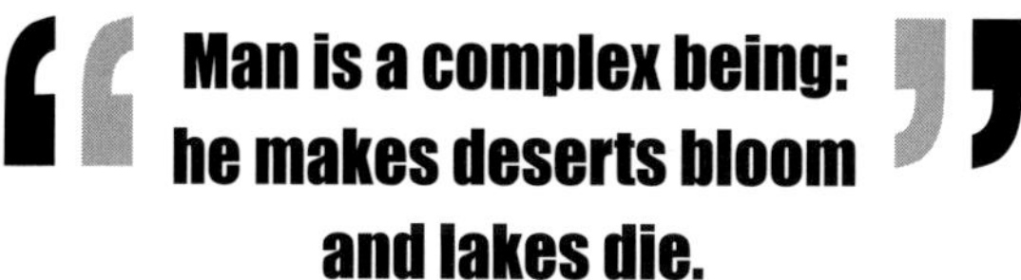

~Gil Stern

Reuse — Pass it forward

What it means to reuse

Reuse should not be confused with recycle. Reuse means using an item more than once, for the same purpose, without reprocessing it. Reuse as a process could also be described as "when the item is used again and again".

In many ways reusing is better than recycling. The most important difference between the two is that reuse prolongs or increases the lifespan of an article. Recycling means converting or turning a product into something new. We need to find ways of reusing things as much as possible.

In India, over the last few years, we have rapidly and carelessly embraced the disposal culture of the West. We are increasingly buying disposable paper napkins, plates, cups, and a whole range of "throwaways". While they come in handy in the immediate term these products cause harm in the long run. The handkerchief is a common, everyday example of reusing things and being a good environmental

citizen. Unfortunately that practice is fast vanishing as we shift to disposable tissues, egged on by marketing advertisements.

Another famous example of reusing would be the practice of re-treading aging tyres. You get more mileage; save energy *and* resources in producing a new tyre and there is one less tyre lying on the garbage heap.

The earlier milk bottle delivery system is also a common example of reuse. People would place the empty bottle on the doorstep and the milkman would replace it with a full one every morning. There were no tetra paks or plastic packets to dispose off.

A different practice that we have followed is that of saving and storing up things like glass bottles, tin cans, newspapers, cardboard, plastic containers, and magazines. Then, ever so often, we call in the *raddiwala/kabadiwala* and sell our hoard to him. This practice is as natural to us as breathing. We have been doing it for generations. The colourful and hectic bargaining between grandma or mother and the *raddiwala* is part of our childhood memories.

The interaction gives us a great feeling of satisfaction. And, so it should. The *raddiwala* is our connection to the process of reusing and recycling. This simple process that involves us directly is probably the easiest of the 3R's to execute.

Quite automatically, we reuse many items in our daily lives—be it at home or in the office or elsewhere. We do it because these reusable articles are a convenience for us. Every rupee that you do not spend means that much less garbage in the landfill.

What is the use

There are several personal rewards to reusing, besides the larger environmental good. When you exchange or sell useful things without reprocessing them, it helps save time, energy, and resources. You personally stand to gain money. By reusing you are also reclaiming some of your original costs.

Try to reuse as much as possible instead of throwing things into the garbage bin. Try to find new ways to reuse things in your daily life. However, if you cannot reuse a particular article or object, then offer it to organizations that collect such items and use them.

On a different and higher level, your reusing deeds provide products to people and organizations with scarce resources. Reuse can help raise money for a good cause. Reuse activities present business opportunities and income to people all along the chain.

The economy, too, benefits. In the poorer nations, reusing has a long, positive, and profitable record. Money is the chief motivator for reusing but it should not be the only one. By becoming eco-conscious, we can deliberately and determinedly reduce our individual garbage footprints.

The result, when we decide to reuse an article, instead of buying a new one, is one less item to leave its environmental impact on the earth. We cannot leave the task of environment protection to the governments, industry, and businesses alone.

It will require the combined efforts of these organizations, institutions and us. We need to give them the support and, above all, the incentive to cut back on wastage. We have to work together to avert the looming disaster.

Reuse of plastic waste in road construction

Who would have thought it! Instead of simply being an eyesore and lining the roadsides of our country, plastic waste could actually make better roads. Waste plastic bags, cups and thermocol are being used to improve the quality of road-making materials. Plastics blended with bitumen by heating, is then used to lay roads.

According to the Central Pollution Control Board of India it was found that adding plastic/polymer waste to bitumen:

- made roads twice as strong as the conventional varieties
- prevented water penetration and thus the formation of potholes
- reduced air pollution by eliminating the process of plastic burning
- reduced cost by cutting down the amount of bitumen needed in the mix

Along with the environment there are a great many gains in reusing things. Just to list a few:

- Increasingly scarce resources, like forests, oil and minerals are saved.
- There are savings in energy as well because replacing disposable products with reusable ones cuts the additional quantity that would be needed to be manufactured.
- The disposal of waste and garbage financially burdens the state and the community. By reusing, you are reducing consumption, which, in turn, cuts costs.
- A reusable product is often cheaper to manufacture than the many single use products it replaces. Thus businesses and consumers benefit by the lower costs.
- Some older items are better crafted and more appreciated.

The managing and disposal of waste has a hierarchy. See the following figure to get an idea of how the same may be viewed.

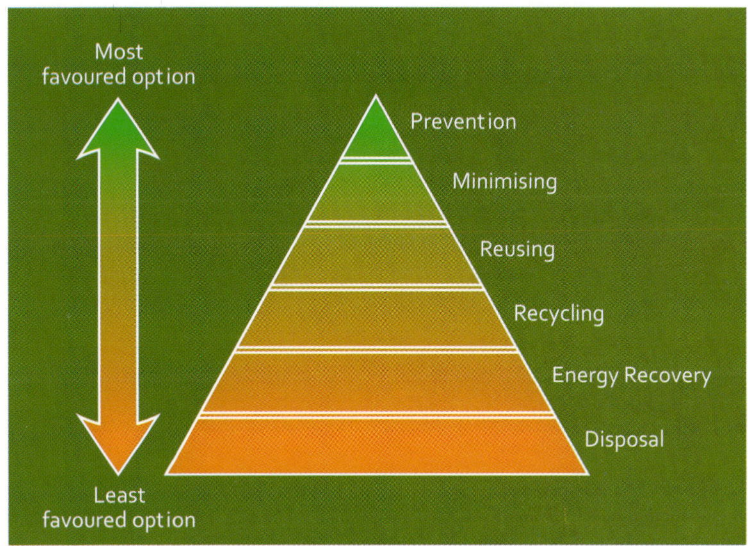

Reuse: a force for good

There are many ways by which we can measure the positive environmental, economic and social impact that reuse has on our communities, towns and villages and the country as a whole.

These include, but are not limited, to:

- Keeping waste out of landfills.
- Reduction of waste and rubbish in landfills.
- Reduction of disposal costs.
- Saving through reduced purchase costs.
- Providing intangible and actual value to the materials donated.
- Profits earned by the seller and the buyer.
- Jobs created or retained.
- Individuals, families, and organizations benefiting from reuse industries.

Reusing clamshell containers

Most plastics in your home can be recycled. But, the berries, biscuits, cookies, and cakes you buy come in plastic containers that cannot be recycled. Here are some ways you can reuse them:

The containers are usually made from polystyrene and cannot be used in recycling plants. However, they can be reused in a variety of creative ways and you never have to send a berry container to the landfill again!

- Use them to pack lunch for school or office. They are also great boxes to put napkins, sandwiches or vegetables.
- Use them to store screws, nails, crayons, ribbons, pictures, decorations, cosmetics, tweezers, nail clippers, etc.

Just use it again—a reuse programme

Reuse: Don't just chuck it away into the garbage bin. It could be very useful to someone else. One man's trash could well be another man's treasure!

It is estimated that an average American uses more than half a million kilos of rocks, minerals and metals during his or her lifetime. These include 365 kilos of lead, 340 kilos of zinc, 690 kilos of copper, 1,636 kilos of aluminium, 14,600 kilos of iron, 1,300 kilos of clay, 12,727 kilos of salt, and 454,500 kilos of sand, stone, gravel, and cement.

Reuse generally takes three forms. You could reuse an item yourself. You could exchange or pass it on to someone else either by donating or just gifting it. The third is by selling it to suitable companies or people.

Or you could . . . .

Start the practice of leaving a book in a public place to be picked up and read by others, who would then do likewise.

You can put the books you have finished reading on the Amazon website or other websites for free. You could sell your books to a second-hand store or donate it to the local library.

> **Did you know?**
>
> Cardboard decomposes and is beneficial to garden soil. So turn your cardboard containers into compost.

You can donate your old computer or its parts to an NGO (non-government organization). They could reuse it to help economically disadvantaged people. Many companies have taken this route. And their benefits and gains have jumped beyond the realm of "earning goodwill in the community".

> ## For 200 years, we've been conquering Nature.
> ## Now we're beating it to death.
>
> ~~ Tom McMillan, to Francesca Lyman,
> *The Greenhouse Trap*, 1990

You could donate your clothes before they wear out and children's toys to Oxfam, SOS Children's Village, the Salvation Army, the Red Cross or any local charity organization.

Instead of dumping your coffee grounds and tea leaves into the trash can, collect them. Add them to the leaves, grass, and vegetable leftovers and even eggshells, and compost the whole lot. They are excellent things to reuse and make excellent compost.

> ## Beware !!
>
> Use caution when storing food in plastic containers. Certain types of plastic can leach into food. For long-term storage or microwaving, it is safer to use glass. It is the best reusable item on the list.

Don't just throw away paper that has been used once. Use both sides and you may also cut them into neat portions and turn them into notepads by using the reverse side. Another good use for used paper is to give them to children to be used for drawing or scribbling.

Buy reusable diapers and nappies instead of disposable ones. You also save money in the long run.

> No one knows just how long glass takes to break down but it is so long that glass made in the Middle East over 3000 years ago can still be found today.

Take reusable bags made of canvas or jute for grocery buying. Pack your lunch in reusable containers.

Pack your lunch in reusable containers, whether it is for school or work.

Preserve wrapping paper from the gifts you receive and reuse them when it is your turn to gift others.

Used clothes can be sold or exchanged or bartered with second-hand salesmen or shops. These can be reused by selling or giving them to new owners.

Donate your old magazines to waiting rooms and local libraries. Printer ink cartridges can be reused. They can be refilled, or resold to the companies that manufactured them. The companies then refill the cartridges, which are then sold back to consumers. Toner cartridges are recycled the same way as ink cartridges, using "toner" instead of ink. This method is highly efficient as there is no energy spent on melting and remaking cartridges. In addition, there are fewer cartridges dumped.

You could check with your local environmental organization on what and how to reuse various articles and household items.

~Jacques Barzun, *The House of Intellect*, 1959

Reusing glass for decoration

Coloured glass can be used in all sorts of decorative ways. One way you could use it is in a water feature, or as part of a wall. People have been known to embed pieces of glass in cement while it's wet (for example in a new wall) to create an interesting effect.

Some people with a talent for craft-making, use glass to make jewellery, lamps and oil burners. There are some tutorials on how to do this online. It goes without saying that you need to be extremely careful if you are breaking or cutting glass for reuse.

Wine bottles make great vases. You can steam the labels off and add flowers for a cosy Italian restaurant look. You could also get even more creative and get some glass paint so that you can decorate the bottles and vases. A hand-painted bottle vase would make a thoughtful gift for anyone who appreciates art.

[To get more on how you can recycle everyday items log on to...greenusesforwaste.co.uk]

How "Reuse" is really useful

- Reuse keeps unusable materials out of the garbage system
- Reuse reduces the amount of energy used to manufacture and even recycle products
- Reuse eases the strain on valuable resources, such as fuel, forests and water supplies, and helps safeguard wildlife habitats
- Reuse creates less air and water pollution when compared to making a new item or recycling
- Reuse results in less hazardous waste
- Reuse saves money in disposal costs

Lifestyle change is a must

The practice of reuse is very high in developing nations. People in these countries reuse a lot more than those in developed countries where higher incomes and better living standards have led to increased demands for goods and disposable products. This means that, with affluence, people stopped reusing and their habits changed.

As we in India reap the numerous benefits of our growing economy we too are falling prey to the ills and bad habits of the developed countries. This is one case where imitation of successful societies is not a good thing. Rather, we should be learning from their mistakes.

Fortunately a rising awareness of the environmental effects is slowly (and hopefully) changing our current habits. There are a growing number of laws that regulate packaging. This is an effort to turn around our harmful past and change the present situation.

So many things can be reused. For instance one can reuse paper by printing on both sides; repair shoes instead of going out and buying new ones merely because there is a crack in the sole. We should always try and postpone the time when an object becomes truly useless to us.

Sadly this change, of attitude and thinking, is happening all too slowly.

5

Recycle — it's the circle of life

The practice of recycling has been a part of human history in one form or another. Long before it became an environmental concern, recycling was a matter of common sense, good economics and a convenient way to create items of basic necessity.

After reduce and reuse, recycle is the third and probably, the most widely practised element of waste management.

Essentially, recycling is the process of converting used or waste items into new merchandise. The intention is to take old stuff and create new goods or a fresh supply of the same goods. The process involves taking materials and items headed for the scrap heap and turning them into goods and products that are valuable again.

The recycling process starts off with the collection and separation of materials like glass, metal, plastics, and paper. These are then sent to the right recycling facilities where they are processed and converted into new products. Many recycled items, like soft drink bottles or aluminium beverage cans, can be reused for their original purpose.

The good old ways

Remember those chilly winter nights when you snuggled under the warm and comforting patchwork quilt that your mom tucked around you? Some of the best quilts and comforters were made from old clothes, blankets, and other materials that our parents hoarded away in the good old days.

Also called *kantha* and commonly found in rural Bengal these quilts are a very good example of recycling. It also demonstrates how ingenious and resourceful our parents were. In times gone by people did not think of recycling in the way we do today. It was always considered sensible and the rich and the poor, alike, did it.

Since the Bronze and Iron ages, we have been recycling metal. Broken and worn out weapons and tools were melted down and converted into useful everyday items. From ancient times, people in the UK have been using wood and coal ash as the crucial element in the making of bricks. The good news is that the practice continues to this day.

Queenly Decree

In 1588, Queen Elizabeth I, of England granted special privileges for the collection of rags that would be recycled to make paper.
[Source www.paper.org.uk/information/]

In the old days when clothes were worn out, they were converted into dish-rags, mops, and quilts. Today, we just throw away our used clothes, adding to waste generation and creating a disposal problem. Throughout history, recycling was driven by basic needs, and supply and demand.

Today, the emphasis has shifted. Now it is not just a matter of sustainable living. The problems of waste and its management impact areas of health—land, air, and water pollution. Plant and animal life are also threatened by the presence and unchecked growth of garbage landfills.

> **Did you know?**
>
> That the first Earth Day was celebrated in 1970.
> [Source: http://earthday.envirolink.org/history.html]

A brief history of recycling

Recycling has been a common practice for most of human history. With just a few examples, we can see how innovative our ancestors were. It is also very indicative of just how ancient the problem of waste disposal is.

2000 BC: Composting was widely practiced in China during the European Bronze Age bronze scrap recovery systems were in place.

400 BC: There are records of recycling collectors going as far back as the times of Plato and Socrates.

1500s: Spanish copper mines use scrap iron for cementation (a metallurgical process of heating and bonding) of copper, a recycling practice that survives to this day.

1800s: "Dustmen" collected the ash from coal fires. Over three and a half million tonnes of coal was burned in London in a year! The dust was taken to dust-yards where "sieving the brieze" or coarse selection of the dust was carried out. This was used as a soil conditioner and for brick-making. The practice is still alive today.

Early 1800s: Many people lived by selling what they could find in other peoples' rubbish; even dog droppings, which was valuable as it was used by tanners for purifying leather.

1813: Benjamin Law developed the process of turning rags into "shoddy" and "mungo" wool in Batley, Yorkshire, England. The process combined recycled fibres with virgin wool.

1890: The British Paper Company was established, specifically, to make paper and board from recycled materials. Waste paper is still obtained from organisations such as the Salvation Army and rag-and-bone men.

Early 1970s: Woodbury, New Jersey, was the first city in the US to make recycling mandatory. Rose Rowan came up with the idea in the early 1970s, of towing a "recycling" trailer behind a waste management vehicle to collect trash and recyclable material. Other towns and cities soon followed suit, and today many cities in the US make recycling compulsory.

It may seem that we have been battling the waste problem for a long, long time. Even today the situation remains critical.

What changed?

Man started to settle down and in a relatively short span of time gave up the nomadic, roving lifestyle. People began to live together in

communities, then villages, which grew to become towns and then cities. This meant that waste began to accumulate in one place and the old "leave-it-behind" practice changed.

Gradually, humankind started to produce waste that did not immediately integrate into the environment. It did not take long for the problem to assume threatening dimensions. The process took many centuries but along with the changes in living patterns came the rubbish and waste problem. As the size of the population and cities grew, so did the waste problem and the need to dispose it.

The huge growth in population and the consequential growth in demand led to the explosion in manufacture. The industrial growth was directly proportional to the growth in demand. As manufacturing technologies improved and things became cheaper and easier to manufacture, our ideas about saving and maintenance, too, changed.

In many cities around the world, pigs were (and stiil are) often used as an efficient method of disposing of municipal waste. In some areas, they still are.

Today, it is so much easier to dispose off "unwanted" items. Not very long ago, people would save every part of every article. We rapidly lost that incentive. Today, we need to regain that motivation. Even better, we should be thinking about *using less* in the first instance. The stakes are so much higher today than what it was fifty years ago.

In 2005, recycling in the USA diverted 79 million tonnes of waste material from landfills and incinerators. That was a rise from the 34 million tonnes in 1990.

[Source: www.epa.gov/epawaste/nonhaz/ municipal/index.htm]

When we recycle, we must understand that our actions have a worldwide impact. To us it may seem a small thing that we do. However, the gains are immense and closely related to our way of living and quality of life. Every little bit counts.

Where have all the trees gone?

According to the World Resources Institute as a consequence of our voracious appetite more than 80 per cent of the earth's natural forests have already been destroyed. Up to 90 per cent of West Africa's coastal rainforests have disappeared since 1900. Brazil and Indonesia, which contain the world's two largest surviving rainforest regions, are being stripped at an alarming rate by logging, fires, and land clearing for agriculture and cattle grazing.

Those recycling seventies

In the 1970s, rising energy costs drove the necessity of more recycling from the general population in the US. Their efforts today are still not whole-hearted.

[Anne Clarke, EzineArticles.com]

Earth Day

22nd April is celebrated as the Earth Day. The day is intended to remind us to appreciate Mother Earth. The name and concept of Earth Day was allegedly pioneered by John McConnell in 1969 at a UNESCO Conference in San Francisco. The first Proclamation of Earth Day was by the City of San Francisco and was first observed in San Francisco and other cities on March 21, 1970, the first day of Spring.

This day was later recognized in a Proclamation signed by the UN Secretary General U Thant at the United Nations after which it is observed each year. At the same time a separate Earth Day was founded by the United States Senator Gaylord Nelson that was held on April 22, 1970. Earth Day is now coordinated globally by the "Earth Day Network" and is celebrated in more than 175 countries every year. In 2009, the United Nations designated April 22 as the International Mother Earth Day.

Good reasons to recycle

There are many good reasons to recycle. The impact and value of recycling can be felt at the individual, community, national, and global levels. Here are some very good reasons for recycling.

Recycling is one of the best ways to reduce the need and demand for virgin resources. It will ensure that we save and protect the natural wealth and resources of the earth. Recycling will also ensure that there is something left for our children.

When we think of recycling, we automatically envision vast mountains of garbage that dot the landscape. Then there is the stench that emanates from these. These are the obvious, in-your-face results of waste and garbage generation. There are other more pressing reasons to recycle and erase the mountains of garbage and the stench.

These are, however, the long-lasting and more dangerous consequences of waste piling up. Waste materials gradually leach harmful chemicals into the ground and release many greenhouse gases into the atmosphere. By recycling we prevent (or, at the very least, reduce) the damaging impact of waste generation on the environment.

What happens to garbage?

It is difficult to get accurate worldwide figures on just how garbage is managed. While large parts of the world do not have reliable intelligence on the magnitude of the problem, many developed countries do have statistical data.

For which reason, we will have to go with the following information and assume the worst for the rest of the world. The trash production in the US has almost trebled since 1960. This trash is handled in various ways. About 32.5 per cent of the trash is recycled or composted, 12.5 per cent is burned, and 55 per cent is buried in landfills.

The amount of trash buried in landfills has doubled since 1960. The US ranks about in the middle of the major countries (United Kingdom, Canada, Germany, France, and Japan) in landfill disposal. The United Kingdom ranks highest, burying about 90% of its solid waste in landfills.

[Source: EPA]

> ### Aluminium anyone?
>
> People in the US throw away enough aluminium every year to rebuild their entire commercial air fleet.
>
> Recycling just one aluminium can, will produce enough energy to run a TV set for three hours!

Another area that is greatly impacted by recycling is energy. It is a well-documented fact that less energy—electricity and heat—is required to recycle material than to produce a brand new product. That, in turn, reduces the cost of the product.

When less energy is required in the manufacture of a product then it follows that there is less air pollution. In a country like India, where over 60 per cent of the energy generated is by burning coal, the reduction in air pollution will be considerable. The savings will also be substantial in terms of costs and coal reserves. Let us not forget the damage that coal mining causes to the earth.

Recycled items used to produce new goods provide a readymade and easily available pool of raw materials for industry. The savings and benefits are incalculable.

Sustainability is another area that recycling affects. Instead of meeting the demand for fresh goods with virgin material, recycling encourages a more efficient supply management and reduced consumption.

Another benefit that recycling provides is in the social sector. It opens up employment avenues to a large number of people. Recycling

requires fairly intensive people power in collection, separation, sorting, and manufacturing. People from low-income groups directly benefit from recycling.

Recycling is the biggest factor in the reduced need for a large number of landfills. Very few people realize that recycling reduces the amount of waste that we have to dispose. The demand for polluting incinerators will also come down. The impact of recycling will be felt *only* when we *all* do it. Educate people around you on the *need* to recycle.

One of the long-term outcomes of recycling will be in the area of technology. Recycling has already inspired the rise and growth of green technologies. Technologies that will emit less greenhouse gases, be more energy efficient, less polluting, and be more environment-friendly should be the order of the day.

That dream day will dawn only if you and I clean up our acts. If not, we doom ourselves to a frightening present, and our children to a bleak future.

There is more to recycling

If you believe that recycling, means only turning old and used stuff into new, think again. There is more to recycling than sending your used and old items to the recycling facility. To make recycling a success and we must take an all-round, holistic course of action and buy recycled products or products that use recycled packaging.

Whenever and wherever there is demand, there will always be someone who will find a way to satisfy that need. It is the same while buying recycled goods as well. By purchasing recycled goods and products, we demonstrate that there is a demand for the same. Thereby an incentive to collect, manufacture, and market these products is created.

Buying products from recycled materials has the benefit of saving scarce resources for the generations to come.

A school in Suphanburi has set up "Garbage Bank" to recycle and reduce garbage from the school. It won an environmental award from the Thai Ministry of Natural Resources and Environment in 2008.

Thammachote Suksalai, a school in Suphanburi, which has about 2,000 students, produces a large amount of garbage. To tackle the problem the school set up a garbage bank two years ago to encourage students to recycle garbage, by trading it at the garbage bank for cash.

Every day, students collect plastic bottles, plastic bags, paper glasses, paper boxes, and newspapers to trade at the garbage bank. Their schoolmates' garbage bank buys the garbage and sells it to garbage collectors. Some of the garbage is used to make utensils for sale at a market. The school has earned almost 100,000 baht so far, money that was used to arrange activities at the school and to fund scholarships for poor students.

Parichat Chansupa, a student, said, "We select garbage for recycling. Students work together to fight global warming and to reduce pollution. Our school also urges students to use both sides of paper."

[Source: www.thaivisa.com]

Did you know?

That paper recycling in Britain began way back in 1921

Believe it or not!

Recycling programmes were established during World Wars I and II and, in many other countries, it continued after the second World War ended. This was only logical, especially for countries that depended on other countries for resources in trade. Nations without an abundance of certain natural resources, such as Japan, continued their recycling practices in full force. They had learned more than one lesson from the terrible war.

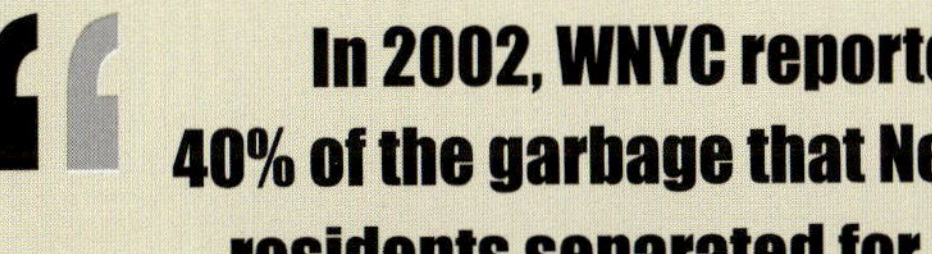

In 2002, WNYC reported that 40% of the garbage that New York City residents separated for recycling actually ended up in landfills.

~WNYC is a non-profit, public radio station in New York City

Recycling innovation: Landfill golf courses

What if you could recycle an entire landfill, filled with millions of tonnes of garbage? This has been accomplished in many places, where the landfill is capped with earth, planted with vegetation and turned into a golf course. Mountain Gate Country Club near Los Angeles is just one such example

Another way to recycle landfills involves capturing methane gas let off by decomposing garbage and using it to generate energy. Another is reusing old landfill pit—where all the garbage was decomposed—by filling it with garbage again.

[Source: Brownfield Golf]

So what is recycling?

First of all, we have to remember that not everything can be recycled, despite our best intentions and efforts. However, glass, paper, metal, plastic, textiles, and electronics are notable examples of things that can be recycled. These can be sent off to collection centres where they can then be sorted, cleaned, and then distributed to appropriate manufacturing locations.

The practice of recycling, over the years, has changed in people's perception. Recycling old paper into new; old clothes into new ones; composting leftover food or garden waste, are the "commonest" examples of recycling.

In an ideal world, recycling would mean producing an endless supply of fresh products from the same material. For instance, making paper from used paper. However, in reality, it is sometimes too costly or impractical to do this. So, in the case of used paper, one of the recycling options is to produce cardboard.

Still keeping the used paper example in mind, several other problems arise. For instance, recycled paper often contains ink residue. It is made up of shorter fibres than virgin paper (made from wood pulp). This means that paper made from recycled paper is not good for some uses, such as in a photocopy machine.

Downcycling

"Downcycling" is a term that came about in the context of recycling. When a recycled product does not match with the quality of the initial product; when the characteristics do not match with those of the original, it is termed "downcycling". It is rare for a recycled product to be exactly the same as the original material from which it was recycled.

One cannot keep recycling a product indefinitely. After a few runs the recycled material becomes unusable. In some cases when recycled goods are no longer usable for the original purpose, their usage can be shifted to another area. There are companies that make furniture and other decorative items out of old newspapers and aluminium cans.

[Source: Stovell Design]

The marvellous thing about recycling is that, with a little creative thinking, it works at several levels. From, the actions of individuals like you and me, to the national and even global scale. It can be as simple as putting new life into old car tyres by converting them into plant containers. It can also get as elaborate as collecting, separating, transporting, processing, manufacturing, and then distributing.

Recycling can take many forms. On a small scale, anytime you find a new use for something old, you are recycling. At a personal level, recycling can mean a cleaner street or neighbourhood.

Recycling can also take other forms. Components, parts or materials from one product can be removed and reused in another product. A good example of this method of recycling is gold. Gold is used in many computer parts. It can be removed, reworked, and reused in new computer parts or other products. Mercury is another commonly used and recycled element.

Whatever form recycling may take, it is very important for us to practise it. Why? Among other things, recycling relates to and affects the beauty and health of our lovely planet. Recycling is one of the most important links in the chain that can make our world sustainable.

Recycling streams

Recycling on the big stage is a different game than what you and I do. For starters, the waste has to be collected. Over a period of time, three different collection streams have evolved. They are a mix of public convenience, government capability, and expense. These streams are "drop-off centres", "buy-back centres", and "collection points". The precise workings of these processes vary across the world.

Drop-off centres

This method requires the waste items to be taken to a specified location. These locations are either fixed or mobile. It could even be the reprocessing facility. This is the easiest and simplest collection system. It, however, suffers from erratic supplies since it is dependent on voluntary contributions.

Buy-back centres

These centres are different in that there is a clear economic/financial incentive for stakeholders to deliver. It also ensures a steady and constant supply. Buy-back centres operate on the profit principle. Once the recycled materials are cleaned and processed, they are then sold for a profit.

Collection points

This form of collection is probably the most varied and takes different shapes in different parts of the world. Trucks routinely collect waste from every street, street corners or designated dustbins. This is sometimes referred to as curb side collection. This system picks up unsorted and mixed wastedegradable and non-degradable.

Sorting

After the collection, the next step in the recycling process is sorting. The mixed waste must be sorted into its different components. In the developed countries, this is done by machine and is very often automated. In the developing world, this is done mostly by hand. The recyclable waste is sorted into three general categoriesmetals, glass, and plastic.

How some recyclables are processed

Iron and steel
The easiest and most commonly recycled materials are iron and steel. They are first magnetically separated from other waste materials and then melted by either an electric arc furnace or a normal furnace. The great thing about steel is that it can be repeatedly recycled into first-rate material with no drop in quality or "downcycling".

Aluminium
Another commonly and widely recycled metal is aluminium. Just like iron and steel, it can be recycled infinitely without loss of quality. The metal is first shredded, and then crushed into pellets. The next step is the melting of pellets. It is now no different from virgin aluminium and is processed the same way as the virgin metal.

There is another great aspect to using recycled aluminium. It requires only 5 per cent of the energy cost of processing new aluminium. This is because the temperature necessary for melting recycled, aluminium is 600°C, while to extract mined aluminium from its ore requires 900°C. To reach this higher temperature, much more energy is needed, leading to the high environmental benefits of aluminium recycling.

Glass
After collection, glass is sorted into colour categories—clear, green, and amber. The glass is then checked for purity. The impurities are removed and the glass is then crushed and dumped into a furnace and melted. The molten glass is then blown or moulded into new jars and bottles.

Processed glass is also used in road-laying materials. Glass makes up 30 per cent of "glassphalt". Glass does not deteriorate and shares the same virtues as iron, steel, and aluminium. It can be recycled ad *infinitum*.

Concrete

Small pieces of concrete can be reused as gravel for construction projects or mixed with other road-building materials. Recycled concrete is also used in the making of new concrete after the impurities are removed.

Batteries

Batteries are highly dangerous and toxic products. They are high on the list of disposal concerns. The potential environmental danger that batteries pose has given rise to specific laws and regulations regarding their disposal.

They need to be handled with great care. Batteries also come in myriad shapes, sizes, and types. This adds to the difficulty in their recycling because each type has to be processed differently. Batteries contain lead-acid, mercury, cadmium, and other toxic materials.

The easiest to handle, are probably lead-acid batteries. These are commonly used in cars.

Kitchen and garden waste

Recycling kitchen and garden waste is the most familiar and easy of all recycling processes. The commonest method of recycling this bio-waste is composting. Various bacteria break down the waste vegetables and leaves  into fertile material, which is then used to enrich the soil. Most of composting is done by people like you and I, on a small scale. Earthworms in a wet compost heap quickens the process.

Clothing

Another popular homely recycling activity involves clothes. We either donate clothes to NGOs like Oxfam and other charities, or turn them into other articles of use, like patchwork quilts, dusters, etc.

Paper

Paper recycling is another popular and widespread activity that is undertaken at both an individual level and industrial scale.

The recycling process for used paper requires it to be pulped and then mixed with virgin (fresh, new) wood. Recycling paper is a case of diminishing returns. The process of recycling used paper breaks down the fibres. Every time the fibres are broken down, their quality diminishes and the resulting paper has to be "downcycled" into inferior products like cardboard or wrapping material.

Nearly all paper can be recycled except those varieties that are coated with plastic or aluminium foil. Papers that are waxed or gummed are difficult to recycle. Gift-wrapping paper is not recycled because the quality is already of a poor standard and unfit for recycling.

Plastics

Plastics are the biggest threat to our environment. Most plastics never degrade or take at least hundreds of years to do so. Recycling plastic is not as easy or simple as recycling iron or steel. The major reason is the enormous number and variety of plastics that exist in the market.

Because of different make-up, plastics have to be sorted first. This is an expensive activity in itself because there is no easy or automated way to do it.

Wood

This is the most satisfying and accepted material to recycle. Because of its obvious and close connection to the environment, people believe that by recycling wood or buying recycled wood they are directly saving the environment. To give this perception a boost, Greenpeace endorsed recycled timber as being an environment-friendly product.

Is recycling worth it?

There are many objections and objectors to recycling. The central arguments revolve around the cost and the economic benefits. However, there are arguments that are far more significant and vital in favour of recycling.

The following table gives a brief overview of the environmental impact and savings that recycling brings about.

Material	Energy Savings	Air Pollution Savings
Aluminium	95%	95%
Cardboard	24%	-
Glass	5%-30%	-
Paper	40%	73%
Plastics	70%	-
Steel	60%	-

While some of the figures may seem negligible as percentages, we need to keep in mind that on a global scale these savings are huge and potentially life-saving.

Is recycling cost-effective?

It is, perhaps, inevitable that the merits and benefits of recycling would create doubt, disbelief, and debate. Besides the environmental arguments nowhere is the debate fiercer than in the economic field.

There are question marks over the cost of recycling and whether its economic benefits are really cost-effective as it is made out to be. While the subject is too vast to explore here, we give a few arguments in favour of recycling.

The governing bodies and municipal councils of towns, cities, and boroughs are in favour of the practice because recycling reduces landfill costs. Without doubt, the worst areas of waste generation are the densely populated areas in and around cities. It should come as no surprise that recycling works best here. It is also more economically feasible and practical because of the economies of scale that are involved.

Economists have an expression for hidden benefits and costs. The term is "externalities". These benefits and costs are not really a visible part of any transaction but are real nonetheless. In the case of recycling, the externalities embrace decreased air pollution and greenhouse gases from burning waste. It takes into account the reduction of hazardous waste leaching into the ground from landfills.

Then there is reduced energy consumption from reduced incineration and manufacturing of recycled goods. There is reduced waste piling up in the landfills. All this leads, directly and indirectly, to a reduction in environmentally damaging mining and timber activity.

Endorsements for recycling

The United States Environmental Protection Agency (EPA) has spoken in favour of recycling, saying that recycling efforts reduced the country's carbon emissions by a net 49 million tonnes in 2005.

In the United Kingdom, the Waste and Resources Action Programme stated that Great Britain's recycling efforts reduce CO_2 emissions by 10-15 million tonnes a year.

A study conducted by the Technical University of Denmark found that in 83 per cent of cases, recycling is the most efficient method to dispose of household waste.

Where do we go from here?

Despite a long and somewhat fragile history, recycling has become a very important part of our lives. In the early and middle 20th century, people were forced to recycle because of scarcity and poverty. People learned to use every bit of an item.

Today that emphasis has changed. We are now being forced to reuse whatever little we have. This distressing situation has come about because today we waste and dispose more than at any other time in history. We have witnessed the Space Age and the Information Age but the 20th century will probably be remembered as the "Waste Age".

We all must think twice before wasting anything, and learn to do with *less*.

Recycling — Myths and Criticisms

This section intends to give a view of what recycling, reducing, and reusing means on a large scale. It will provide you with a glimpse of how small actions by every one of us can add up to make a considerable difference.

Any discussion on recycling invariably attains fever a pitch. There are many who argue that there are no benefits of recycling. The debate on recycling has moved from waste disposal management to economics, sustainability, employment, social responsibility, resource management, government and public responsibility, and health.

The most important issue, of course, is the protection of the environment. It may be stretching it a bit too far to say that the impact of recycling could ultimately affect our survival. However, what is increasingly clear is that recycling has become a crucial cog in the sustainability mechanism.

For most of us, the concept of recycling has individual and local implications. Through our individual recycling efforts, we are able to reduce our expenses. The local garbage bins are no longer overflowing and smelly.

There are obviously many other larger implications which we fail to recognise. Here, we will try to explain and answer the recycling question. We will attempt to clear any doubts you may have on the subject.

In Europe, Austria heads the EU in its recycling efforts with approximately 60 per cent of its waste being recycled.

[www.environment-green.com/http://www.friendsof europe.org Friends of Europe Green Week Report,thegreenmom.blogspot.com/]

Recycling is a waste of resources

... and not worth the effort and time! This is one of the loudest arguments against recycling. The critics of recycling also argue that if recycling really works it will be reflected in the behaviour of the market and that the prices of depleting natural resources would rise, they say. That is not happening they argue.

Critics further argue that despite the increase in consumption of fossil fuels, supply has not so far diminished to that dangerous level. In fact,

cynics argue that reserves of oil and other fossil fuels are growing as we make more and more discoveries and there is actually no need to panic. This is, however, a hollow argument. It is true that we are finding and exploiting newer resources of oil and natural gas. However, most of these "discoveries" were actually made years ago. These are being mined now and extracted because the older reserves are rapidly running out.

A not-so-obvious benefit of recycling is its potential to reduce our consumption of fossil fuels, because oil needs to be refined in order to create plastic resins for products such as soda bottles, food wrappers, and throwaway cameras.

It is baseless to assume and even argue that our natural resources are infinite. And the rate at which the per capita global consumption is rising, particularly in large countries like India, China or Brazil, we will soon gobble up all our remaining resources.

According to statistics given by the British Petroleum, the reserves/production (R/P) ratio of the world's fossil resources has been estimated as shown:

Fuel	R/P Ratio
Oil	40 years
Natural Gas	62 years
Coal	224 years

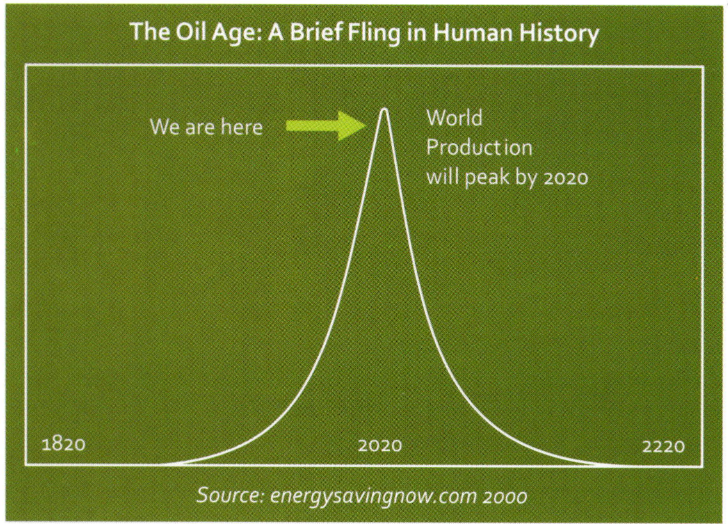

Recycling doubters argue that "Innovation" has increased the availability of resources. Thanks to innovation, we now produce about twice as much output per unit of energy as we did 50 years ago and five times as much as we did 200 years ago.

Optical fibres carry 625 times more calls than the copper wire in use 20 years ago. Bridges are now built with less steel, and automobile and truck engines consume less fuel per unit of work performed . . . the list is endless. Human innovation continues to increase the amount of resources at our command.

Necessity is the mother of invention. This proverb has been proven time and again. It is the dwindling of resources that push us to be more innovative. This is evident from the many innovative ways that we are recycling waste materials today.

It should be mentioned here that much of the information and "facts" used in these arguments are taken from studies that are sponsored by

people and institutions that have vested and financial interests in opposing recycling. Their "arguments" favour certain industries that are hurt / hit by recycling.

Mills and recycled paper

The amount of recovered paper used by paper mills as a proportion of their total output is known as the "Utilization Rate". The rate for the UK is 72 per cent, one of the highest in the world.
[CPI - Competitiveness Study for Paper Related Industries in the UK, December 2000 www.letsrecycle.com]

Recycling quantities are minuscule

Critics of recycling state that no matter what we do, we can only recycle between 25 per cent and 30 per cent of solid waste.

As the infrastructure and facilities for recycling and composting expand, the overall savings will also improve. Today, nearly 50% of solid waste materials are recycled. In the US alone, the rate of recycling doubled between 1980 and 1990. From just 9 per cent it spiralled to 17 per cent. And by 1995 it had risen to almost 27 per cent.

Today that figure is between 30 per cent and 50 per cent, with a potential for even more. As recycling programmes around the world grow and become more and more efficient, there will be striking increases in the amount of recycled materials as well.

In the US, 95 per cent of all waste plastics and two-thirds of all waste paper still go un-recycled. However, the recycling industry has still been steadily maturing as it now diverts almost 24 per cent of the nation's municipal waste into productive uses.

In 1995, 62 per cent of the 100 billion aluminium cans produced annually were returned for recycling.

The recycling of food, textiles, and construction materials is on the rise. We are finding more and more ways to reuse these items. There are studies also which demonstrate that more than 60 per cent of these substances can be recycled efficiently.

"Where there is a will, we will find the way!"

Recycling is an expensive affair

Detractors of recycling argue that recycling is costlier than garbage collection and disposal. This was probably the most reasonable and persuasive argument of all. There was a time, not so very long ago, that it was definitely cheaper to just dump waste material into landfills. Like all things, there were initial teething troubles. We had to learn, adapt, and improvise to get the right technique.

The cost of recycling is very often compared to the cost of disposal. This comparison puts recycling in a very poor light. What opponents forget (sometimes conveniently) is that the cost of collection and disposal should be matched with the cost of collection and recycling. This

approach to the evaluation of the two systems shows that recycling is actually far more cost-competitive than mere collection and disposal.

Another argument on the cost-effectiveness of recycling is that of the long-term costs involved. For many recycling is still an additional expense. But, in reality, recycling is the primary target of waste disposal and not supplementary. Recycling becomes economical and a generator of employment and profitable avenue. More and more public service administrators are realizing that when conventional collection is married to recycling and composting, it makes better economic sense.

As recycling programmes grow along with the amount of materials collected, the economics of scale take over. Collection and recycling becomes more cost-effective and less expensive than conventional collection and disposal.

It is a mountain but not as we know it!

Mount Rumpke is the highest point in Ohio, USA, at over 1000 ft. It is no Mount Everest; however, Mount Rumpke is literally a mountain of trash and is located in the Rumpke landfill!
[Source: www.environment-green.com/]

When recycling programmes are well designed and executed, these prove to be far more economical than mere trash collection and disposal. Municipalities and communities that get the most out of recycling save money and enhance the quality of their environment as well.

Recycling is one of the best methods of slowing down the consumption of natural resources of all kinds. Reduce, reuse, and recycle as much as possible. Becoming more efficient in our use of energy and resources is the other great way of slowing down the problem of fast depleting natural resources.

A 1996 *New York Times* magazine article noted in an editorial that it actually costs the city less ($10 to $40 per ton) to deliver its wastes to recyclers than it does to dump it on its own Staten Island Fresh Kills landfill ($42 per ton).

Dump it or burn it?

One anti-recycling argument is that the process costs more than having garbage dumps or landfills. The argument runs on the basis that the cost incurred in collecting and dumping, or incineration, at landfills, is much lower than sorting, distributing, and recycling.

The idea that just dumping is more effective than recycling does not take into account the huge damage caused to the environment.

Landfills pollute and there is no question about that. Pollution makes dumping and incineration methods unsafe and risky.

Then there is the space required for creating landfills. There is one rather laughable claim that if all the garbage in the US were placed in one landfill site it would require only 10 square miles. Surely that is not such a large area, runs the argument.

As it stands, landfill capacities are reaching their limits. Many landfill sites have only 10 years of capacity left. Many more have even less than five years to reach capacity. Landfills are a scarce and valuable commodity. Recycling increases the lifespan of landfills and reduces waste disposal costs. Transportation to new far-flung landfill sites will only increase disposal costs.

Also, by expanding existing landfills or creating new ones, communities are affected. No one likes trash dumped on his or her doorstep. This was amply brought out by the demonstrations and social disturbances that occurred in and around the Millennium City of Gurgaon in Haryana. When the civic authorities tried to create new landfills to accommodate the garbage generated by the city, the villagers were deeply hurt and protested loudly against it.

> Plastic not only adds to landfill space, it takes forever to decompose. Used plastic dumped into the sea kills and destroys sea life at an estimated 1,000,000 sea creatures per year!
>
> *[Source: http://www.all-recycling-facts.com/recycling-statistics.html*
> *http://www.mercergroup.com/plastic.htm]*

Landfills do not just have economic or financial implications; they have social implications too. The aim, therefore, should be to lessen the undesirable impact on local communities.

> **I think the environment should be put in the category of our national security. Defence of our resources is just as important as defence abroad. Otherwise, what is there to defend?**
>
> ~Robert Redford, *Yosemite National Park Dedication*, 1985

Burning or shoving waste into incinerators is another argument that is used against recycling. Many cities, municipalities, and local communities have adopted this method of disposing off their waste. A study evaluating Florida's seven largest incinerators revealed that these facilities regularly burn significant amounts of highly recyclable materials.

The toxic output in terms of air pollution and gas emissions are incalculable, even though the incinerators are modern and super efficient. About 30 per cent of incinerated waste left behind is ash.

Moreover, the burnt remains are poisonous. Incinerators produce organic compounds like carbon dioxide, sulphur dioxide, nitrogen oxide, and other acid gases that landfills and recycling methods do not.

> **There must be a reason why some people can afford to live well. They must have worked for it. I only feel angry when I see waste. When I see people throwing away things we could use.**
>
> ~Mother Teresa

Municipal waste concerns mainly revolve around minimizing the amount of garbage created in the first place, and making products last longer.

Landfills and incinerators create jobs

Advocates of the landfill and incinerator method of waste disposal proclaim that these activities create jobs. Yes. This is true. However, the number of jobs created is relatively small.

Recycling and composting generate considerably more jobs than just tipping at landfills or incinerating waste. Recycling creates jobs at the collecting, sorting, and production stages of the system. Its impact can be seen in the rural and urban settings.

It has been found that on a per-tonne basis, sorting alone requires 10 times more jobs than landfills. If we move further down the recycling stream, we will see that making new products from recycled material utilizes the skills of even more people. They even earn higher wages than those involved with landfills.

Paper mills based on recycling and those involved in plastic recycling, employ over 60 times more workers than the landfill projects. The figures are even more heartening when one looks at computer restoration and recycling. This industry generates 68 times the number of jobs than landfills or disposal.

There is another interesting and heartening statistic. If the 25.5 million tons of durable goods now discarded were reclaimed through the types of reuse operations mentioned earlier more than 220,000 new jobs could be created in this industry alone.

> Glass is one of the very few products that can be completely recycled repeatedly. But most often, it ends up in landfills and never decomposes.

Governments and regulations are unnecessary

Many economists and market supporters believe and argue that the private sector, driven by profit incentives, is best placed to carry out recycling. They want governments and other public sector bodies out.

This, unfortunately, is an area where private, profit-based initiatives have faltered badly. Only when it has been through a complex, organic, and evolutionary process involving people, communities, and civic bodies that recycling has taken off. The public sector can shape, regulate and provide a system of checks and balances married to reward and penalize schemes that will check product consumption and reduce disposal. The public sector is actually in a better position to plan and execute recycling programmes. Its very nature and structure will guarantee that the community, in all forms, benefits.

The government and other public sector institutions operate through rules and regulations, agreements, local ordinances. These take the form of taxes, laws, subsidies, business regulations, environmental laws, and land use requirements which, together, ensure that waste management infrastructures and recycling initiatives flourish.

For instance, making manufacturers responsible for the cost of garbage, would compel them to reduce excessive packaging and, by extension, waste.

The public sector and the government are the best agencies to set up systems and establish rules that ensure reduction, reuse, recycling and composting. There should be laws /checks which stop us from dumping our trash in someone else's backyard, reduce pollution, make us more frugal, and create jobs and new businesses. It is these philosophies that have been the reason behind the success of recycling programmes.

Every year in the UK, households throw away waste, which is equivalent of 3 ½ million double-decker buses (almost 30 million tonnes), a queue of which would stretch from London to Sydney (Australia) and back!!

Waste disposal and recycling also have a social and health impact. So it stands to reason that public institutions are best equipped and motivated to drive recycling initiatives.

It is through these agencies that the marketplace operates by enacting bottle bills, minimum recycled-content product standards, landfill disposal bans, mandatory recycling, recycling-related business attraction incentives for the private and community sectors, and procurement programmes for recycled-content products.

These public sector interventions are needed in order to transform solid waste management problems into materials conservation and recovery opportunities. It is only then that the private sector will fall in line and do what they do best—operate efficiently and profitably.

Recycling is a success story

"There are no passengers on *spaceship earth*, only *crew members*," said Marshall McLuhan in 1965. "Whatever happens to the ship also affects us. We would like to keep our spaceship safe."

Recycling is probably one of the best ways to save the planet. Yet many people are reluctant to do the 'right thing'. Recycling is a decision based on caring for the earth, and of having concern for the world around us.

If we do it right, then recycling can be a practical exercise and have a positive impact too. People need to only open their minds to find that the setting up of a recycling system in our daily lives is not a difficult matter. It is not expensive or time consuming to implement either.

One litre of oil can pollute one million litres of fresh drinking water!

Recycling is a sustainable and rewarding activity. It is also a painless undertaking. Recycling in all its forms and shapes can be, and is, a success.

The stories of recycling are varied, interesting, and successful. To give you an idea of its variety and success, we have compiled an album of snapshots taken from around the world.

Perhaps you could be inspired. You too could script a success story!

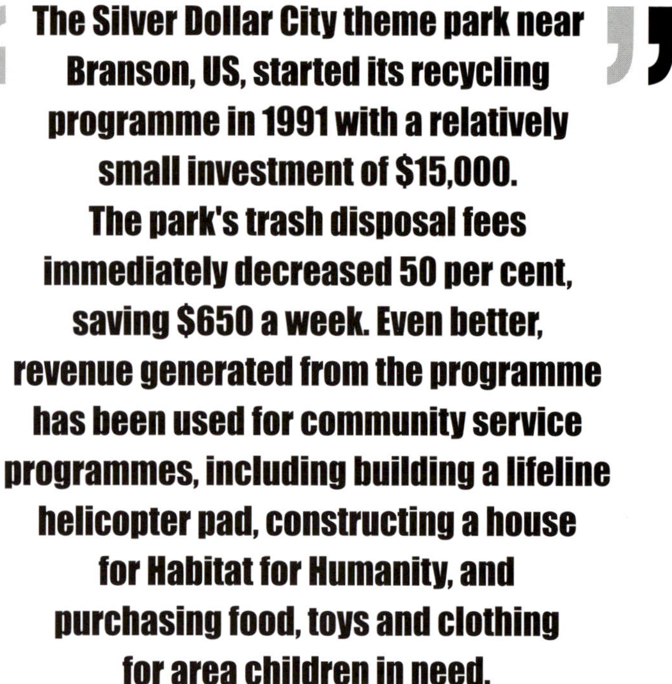

The Silver Dollar City theme park near Branson, US, started its recycling programme in 1991 with a relatively small investment of $15,000. The park's trash disposal fees immediately decreased 50 per cent, saving $650 a week. Even better, revenue generated from the programme has been used for community service programmes, including building a lifeline helicopter pad, constructing a house for Habitat for Humanity, and purchasing food, toys and clothing for area children in need.

~Reported by Becky Mollenkamp

Zero roadside garbage

A group of concerned citizens ordinary folks like you and me - set up the "Clean Ahmedabad Abhiyan Committee" in Gujarat. Several voluntary organisations and the Municipal Corporation of Ahmedabad collaborated with them.

Their goal was to find and implement permanent and sustainable solutions to the problems caused by roadside garbage. The waste and filth rotting by the roadside was causing various health and sanitation problems. The committee wanted to deal with issues of unsanitary solid waste and recycle it.

Through this programme, the Abhiyan Committee hoped to raise public awareness and get people to participate. They also wanted to safeguard public health. The following are some other objectives they planned to achieve:

- Zero garbage on roads
- Minimum landfill, maximum recycling
- Creation of economical benefits for poor through self employment
- Redesign the municipal waste collection system

The Abhiyan Committee started a public campaign to educate people and make them aware of the problems. People and households were taught to separate wet and dry garbage. They were drawn and co-opted into the project.

A special bag with three compartments was developed to segregate and store recyclables paper, plastic, and miscellaneous. The heavy-duty bags are reusable and last two years. Segregated recyclables are collected from each house by "rag collectors" each of whom have been assigned 100 to 200 houses. The increasing multi-storey apartment societies use "community bins" for collection of kitchen waste. A specially designed truck removes and replaces full bins. A low-cost kitchen waste digester and composer have also been developed.

The community initiative was successful beyond expectations.

- They put in place a system of collecting dry and wet waste. The waste is recycled and used as compost.
- A cooperative rag collectors' organisation has been set up and the income they earn has generated tremendous confidence among the collectors and boosted their self-esteem.
- There is improvement of health standards in the community. They are able to develop a low-cost kitchen waste digester.

There were other important lessons learnt as well …

- Voluntary participation brought benefit to all levels of the community.
- Focus is the key. In this case, recycling.
- The media is necessary and plays a very important role in relaying frequent messages on cleanliness.
- Developing low-cost and appropriate technology is an integral part of any such programme.

The entire programme was funded by voluntary contributions. This, probably, was the best lesson of all. The success of this project has improved the living conditions of hundreds of rag- pickers. By removing the garbage from the streets, the "Clean Ahmedabad Abhiyan Committee" has not only changed the city's landscape but also provided the much needed social and health benefits.

And all it took was a change of heart; a sense of ownership for our surrounding and a little bit of caring!

> Recycling just one plastic bottle can save the same amount of energy needed to power a 60-watt light bulb for six hours.

Some heartening examples

- Company A did not think it generated a significant amount of waste, but when the company reviewed its activities and introduced more efficient ways of handling cardboard, waste elimination bills were cut by 55 per cent. The Company also saved staff time, increased staff awareness, and reduced their waste by 577 tonnes (in the first year).

- The managing director of Mounstevens Ltd., a manufacturing and retail baker, increased staff awareness and introduced careful separation of waste. The expected benefits include cutting waste bills in half and saving $8800 and 26 tonnes of waste. "We would have to sell a lot of extra doughnuts to make that sort of impact on our profitability," said the managing director.

- Company B instituted a facility-wide municipal waste recycling programme including metal, cardboard, paper, wood, plastic and glass. More than 50 per cent of the municipal waste generated by the company is now recycled. The programme greatly reduced disposal costs and generated enough revenue from marketing the recyclables which funded the programme's operating expenses, including wages and benefits, equipment operation and maintenance, utility costs, and programme improvements.

[Source: Studies provided by the International Finance Corporation]

US Paper Recycling Facts

- In 2003, Americans recovered 49.3 million tonnes of paper for recycling. This represents 50 per cent of all paper consumed in the US.

- In 2003, recovery of old corrugated containers (OCC) rose to a record high of 75.8 per cent, recovery of old newspapers (ONP) rose to a record high of 73 per cent, and recovery of office paper rose to a record high of 48.3 per cent.

- Americans recycle 270 million pounds of paper every day.

- More than 37 per cent of the raw material used to make new paper products comes from recycled paper.

- Nearly 80 per cent of all US papermakers use recovered fibre to make new paper products.

- Every ton of paper recovered for recycling saves 3.3 cubic yards of landfill space.

- Currently, more paper is recovered for recycling than is land filled. By weight, more paper is recovered from municipal waste streams for recycling than all glass, plastic, and aluminum combined.

[Source: The American Forest & Paper Association (AF&PA)]

Japanese firm opens paper recycling plant in China

Marubeni Corp., of Japan, has announced plans to set up a paper recycling joint venture in China that includes nine paper material wholesalers.

The company decided to open the plant in response to the surging demand for recovered fibre in China.

The joint venture is located in China's Jiangsu province.. The plant will collect, import, and process used paper and also sell the processed material to paper manufacturers along the RiverYangtze.

The venture hopes to achieve an annual volume of 100,000 tonnes.

[Source: 2004 G.I.E. Media, Inc.
2004 Gale Group]

It's not the same old story

There is a saying that "success has many fathers ..." and that is true for recycling. The examples that we are highlighting here show that there are many, many ways to recycle: making a difference comes in many different ways and forms.

As the world becomes more and more aware of the challenges and dangers posed by unchecked waste disposal, many communities, cities, agencies, and ordinary people are finding ways to check this trend. From small recycling enterprises to citywide programmes to individual composting practices, people are rising to the challenges.

Each community, no matter where it is located, faces its own unique tests, challenges, and struggles in dealing with waste. Many have worked out their own approach and set up policies to tackle the problem.

The examples here give lessons learned by others and may provide suggestions when you start your own recycling initiative or community programme.

The amount of electrical waste

Electrical and electronic goods, which make up 4 per cent of European municipal waste, are growing three times faster than any other waste category.

From sewer sludge to campus conservationist

The Chennai Petroleum Corporation has a large refinery and is a heavy consumer of water. It has pioneered a brilliant recycling model. It actually pays the Chennai Municipal Corporation Rs 8 per kilolitre of sewage.

After letting it settle in holding tanks, it uses reverse osmosis to filter out solids. The resulting water is 98.8 per cent pure and good enough for use as process water. The sludge then flows into vermicomposting beds to produce manure that is used to maintain the vast campus, lush and green. The scale of operation is enormous 1,500,000 litres an hour or 40 per cent of the refinery's needs. It's a win-win solution for the city, the business and the environment. The project has generated great interest among other industries. *(Good News India website)*

Citywide rainwater harvesting

Rainwater harvesting and recycling are the current new buzz in Chennai. Residents are cleaning and reusing wash water in toilets and gardens. Alacrity Foundations is one of Chennai's most conscientious builders. They had installed rainwater harvesting (RWH) systems long before the ordinance was even issued. They are now busy installing systems to recycle grey water. Their belief that over 80 per cent of the water that flows into the sewage can be reused, has led to a steady flow of successful recycling stories in Chennai.

Indukanth Ragade, an organic chemist, has taken the lead in executing rainwater harvesting and wastewater management projects in Chennai. As vice chairperson of Alacrity Foundations Private Limited, he started introducing rainwater harvesting in all its housing projects from 1993. So far the company has introduced rainwater harvesting in over 150 projects comprising over 4,500 flats. By combining RWH with wastewater treatment and reuse, Alacrity Foundation, has found a sustainable solution to Chennai's acute water scarcity problem. The proper application of this technique can greatly reduce dependence on external sources.

According to Alacrity's calculations, RWH alone has the potential of meeting about 30–40 percent of the flat complex's annual water needs. This can be further increased to 60 percent by reusing wastewater after *in situ* treatment.

The wastewater is of three kinds:

- About 30–40 per cent of wastewater is from closets for flushing, and cannot be reused
- About 10 per cent of wastewater comes from kitchens. It is not reused, as the level of nutrients is high.
- Only the water used for bathing and washing clothes can be treated and reused for toilet flushing or groundwater recharge. It constitutes 50–60 per cent of the total consumption. For recharging the groundwater, the wastewater is diverted towards a specially prepared soil bed, in which semi-aquatic plants are grown. If the water is to be recycled, then the bottom of the bed is made waterproof to prevent seepage.

From each complex, a network of three different pipes separate wastewater at the first stage. Such projects require moderate capital investment as well as minimal maintenance.

In the 12 localities of Chennai where Alacrity has worked, the system has been operating smoothly. One of them is in Tambaram, an 80-flat apartment block, where the system has been in place for more than three years. Here, the quality of drinking water has remained stable and a dry bore well has begun yielding water again! The system operates on the principle of gravity with no related problems of chemical permeation, smell or mosquito breeding. (*Rainwaterharvesting.org*)

Honeycombe Leisure Plc, in the UK, which runs 32 pubs and bars, implemented a glass recycling scheme, which resulted in sustainable cost savings on the tens of thousands of used bottles annually. Honeycombe Leisure also introduced water regulators to eliminate unnecessary water use and air filters to improve the atmosphere. "All of these initiatives are already making a positive impact both on the profitability of our company and on the environment," said Michael Norris, Honyecombe Leisure's financial director.

[Source: Study provided by the International Finance Corporation]

Did you know that composting can:

- Suppress plant diseases and pests
- Reduce or eliminate the need for chemical fertilizers
- Promote higher yields of agricultural crops
- Facilitate reforestation, wetlands restoration, and habitat revitalisation efforts by amending contaminated, compacted, and marginal soils
- Cost-effectively remediate soils contaminated by hazardous waste
- Remove solids, oil, grease, and heavy metals from storm water runoff
- Capture and destroy 99.6 per cent of industrial volatile organic chemicals (VOCs) in contaminated air
- Provide cost savings of at least 50 per cent. This is over and above conventional soil, water, and air pollution reduction technologies.

UK wood consumption

The UK accounts for about three per cent of current global wood consumption, with only about one per cent of the world's population.

The good old ways

The old ways are still the best ways. RWH techniques from ancient times still work efficiently. It is one of the most important ways to recycle scarce water resources and preserve what little comes down as rain.

Community-based RWH—an ancient concept—still works wonderfully. A survey conducted by CSE of several drought-struck villages found that those that had undertaken RWH and/or watershed development in earlier years had no drinking water problem whatsoever. They even had some water left over to irrigate their crops.

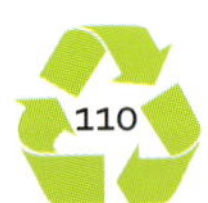

Street dustbins are the problem, not the solution!

Urban India generates well over 40 million tonnes of waste every year, but it is yet to develop a comprehensive safe disposal and recycling system for small, medium, and large towns. Mounting garbage dumps situated in low-lying areas and the outskirts of towns pose serious health and environmental risks. Perhaps we need to be reminded of what Gandhiji said: "For India, sanitation is more important than independence."

Suryapet, a small town in Andhra Pradesh, has earned the rare distinction of being India's first "dustbin free" town.

The sleepy town of Suryapet woke up to a reality that has caught most other towns and cities napping. Although of little historical value and tourist significance, this small town in Nalgonda district, Andhra Pradesh, is way ahead of many urban centres in the country. It has earned the rare distinction of being India's first "dustbin free" town.

Spread over an area of 34 square kilometres, Suryapet supports a population of one lakh and thirty thousand. Some 20,000 people from adjoining villages throng here every day, as the town is an important commercial and educational hub. Needless to say, the town's resident population, as well as its daily visitors, put an enormous strain on its 218 km of drains and 241 km of roads.

"Overflowing dustbins, garbage-littered streets and a stinking town was my inheritance," recalls municipal commissioner S A Khadar Saheb who took up his appointment towards the end of 2002. Like most other towns, Suryapet's annual income of ₹8.45 crore was too small to manage its annual garbage output of 12,000 metric tonnes. Seized of the situation, Khadar Saheb knew he was up against a formidable challenge.

Within days of taking up office, Khadar Saheb took to the streets with his small army of 235 public health workers. He set about clearing the streets of their overflowing dustbins; some 360 overburdened dustbins were soon relieved of their duties! "Only by removing the dustbins could we have thought of better options," quips Khadar Saheb. Instead, each household was provided with two different-coloured bins to put their rubbish.

The aim was to involve the entire community in the transformation. Street meetings, street plays and door-to-door visits marked the social mobilization process. In addition to their routine duties, municipal workers distributed leaflets and encouraged households to segregate their waste at the source. Municipal staff greeted members of every household with traditional *kumkum* (vermillion) and made them feel proud of being part of the effort.

As soon as people in the town began segregating their wet and dry garbage into the green and red bins provided by the municipality, nine tractors, three tricycles, and three cycle-rickshaws were pressed into service, collecting the garbage from homes. It was a treat watching people rush out to unload their bins into the municipal vehicles, at the blow of a whistle!

Unbelievable though it may sound, the municipality spent just ₹ 1.24 crore a year on cleaning up the town. No additional workers were hired, nor were extra resources mobilized. "Transparency and efficiency were nevertheless revived," says Khadar Saheb. Motivational incentives, regular monitoring, and efficient management provided back-up support to the much-needed cleanup effort.

Long before the Supreme Court deadline for the implementation of the Municipal Solid Waste Rules, 2000 came into effect in January 2004, Suryapet had already achieved the rare distinction of being the first town to have segregated its waste at source; to have collected it without littering the streets; and to have composted the same to generate additional income. Segregation of waste helped the municipality earn ₹ 26,000 in a year, just from the sale of r ecyclable plastic.

The Municipal Solid Waste Rules call for source segregation as the first step, before cleaner composting and recycling of urban solid waste. This discourages the creation of hazardous dumping sites, on one hand, and the need for environment-unfriendly waste incinerators on the other. Suryapet has clearly demonstrated that dumping sites and incinerators may become a thing of the past if indeed there is the political will to bring about change.

Khadar Saheb is convinced that Suryapet's success can be easily replicated in towns with a similar population. "Resources are not the constraining factor, good intentions are," he argues.

At a time when most municipalities are experiencing the pressures of meeting the Supreme Court's deadline, Suryapet offers a viable solution to the problem of urban waste. Many municipalities are also considering privatization as an easy option. *(InfoChange News and Features, 01/08/2004)*

Indian cities generate 42 million metric tonnes of waste annually. Of that amount, only four million metric tonnes are retrieved for recycling. Another four million metric tonnes are disposed of in uncontrolled and unmonitored dumps. This is a huge problem that needs to be solved and stopped.

Separation of recyclable waste is not seriously practised in Indian households, shops, and other establishments. At least 15 to 20 per cent of the country's total waste could easily be segregated at its source. This could then be usefully recycled downstream.

Treasure from India's wastelands

Bharath Agro Products is a venture started at the Tiruchirappalli Regional Engineering College, Science & Technology Entrepreneurs Park (TREC-STEP). Bharath Agro Products is a member and grantee of *info* Dev's Global Network of Business Incubators.

Bharath Agro is involved in the manufacturing of a tractor-mounted pulverizer for collecting waste from Julie Flora, a common crop in several Indian states. The Julie Flora (known as *Karu Velam* in Tamil and *Babool* tree in Hindi) is a firewood tree in dry lands across India. After collecting Julie Flora's thicker branches and stems for firewood, farmers are left with waste, such as thorns and small branches that comprise about one-third of the total crop size. This waste is then burned on the farmlands, which causes pollution.

Mr Kannappan, a mechanical engineer, and Mr Paneerselvam, a farmer, decided to conjoin their knowledge and skills together in agricultural production practices and technologies, and start their own company. Rather than leaving Julie Flora waste to burn and cause pollution, they wanted to find an innovative way for collecting and making use of this waste. Kannappan and Paneerselvam then designed and fabricated a

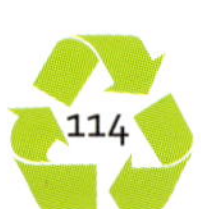

tractor-mounted pulverizer that would efficiently collect the Julie Flora waste and turn it into a product that could be sold to biomass power plants, boiler manufacturers, and even major hotel kitchens. This is a unique product in the Indian market with no direct competition, which naturally offers tremendous market potential to Bharath Agro Products.

The tractor-mounted pulverizer can collect up to five tonnes of Julie Flora cut-waste per day, which translates into collection of almost two-third of available waste. As a result, the innovative pulverizer prevents the potentially enormous damage done by slash-and-fire systems to the fragile environments of India's dry farmlands and provides an additional supply channel for energy producers.

The innovators rightfully claim that, in their resource-poor district alone, these machines have the potential to reclaim waste worth more than ₹ 50 million [approx. $1.1 million] every year.

There are 604 districts in India and rough estimates say that more than 150 districts have Julie Flora waste: the wealth creation potential from harvesting this waste, due to the tractor-mounted pulverizer innovation alone is more than $150 million. In addition to this, it has the potential to create 120,000 jobs when operating at this level. Other benefits such as carbon offsetting to reduce carbon emissions are also significant. TREC-STEP believes that it is only a matter of time before this innovation capitalizes on its potential, resulting in wealth creation, jobs, and environmental protection in rural villages. (*info*Dev website)

Waste Not, Want Not in
Dharavi, Asia's Largest Slum

~By Dan McDougall, *The Observer*, March 4, 2007

Selfridge Air National Guard Base (SANGB) is a joint-service military facility located approximately 20 miles north east of Detroit, Michigan, US. As part of their recent base renovation plan, the Base contracted DEMEX, Inc., to demolish 22 buildings located throughout the base.

In total, about 74 per cent of the demolition debris was recycled or reused, and was diverted from disposal in the landfill. These reuse and recycling practices also resulted in a contract bid price that was 25 per cent below competing estimates, demonstrating significant cost savings to the government.

If you have the patience to look closer, you will find here one of the most inspiring economic models in Asia. Dharavi may be one of the world's largest slums, but it is by far its most prosperous and thriving business centre propelled by thousands of micro-entrepreneurs who have created an invaluable industry, turning around the discarded waste of Mumbai's 19 million citizens. A new estimate by economists of the output of the slum is as impressive as it seems improbable £700 m a year.

For Dharavi's detractors, mainly Mumbai's city fathers and real estate developers, keen to get their hands on the prime land beneath the sprawl, the shanty-town is an embarrassing boil to be lanced from the body of an ambitious city hoping to become the next Shanghai.

But for a growing number of environmental campaigners Dharavi is becoming the green lung stopping Mumbai from choking to death on its own waste.

All along the Apna Street, hundreds of barefoot street children, human recycling machines, scurry back and forward, hauling bundles of waste —plastic, cardboard or glass—retrieved from Mumbai's vast municipal dumps. From every alley come the sounds of hammering, drilling, and soldering. In every shack, dark figures sit waist-deep in piles of car batteries, computer parts, fluorescent lights, ballpoint pens, plastic bags, paper and cardboard boxes and wire hangers, sorting each item for recycling.

Workshops reveal everything from aluminum smelters recycling, drink cans to perspiring bare-chested men stirring huge vats of waste soap retrieved from rubbish tips and local hotels. Walking through Dharavi, home to an estimated 15,000 single-room factories, it becomes difficult to conceive of anything that is not made or recycled here.

In the mud behind a hut employees use metal bars to bash dents out of one-gallon cooking oil cans. In an adjacent courtyard 20 more workers squat on stones and clean and hammer two-litre cooking oil cans back into shape. Along the assembly line, the cans are dipped into scalding water, given a scrub, and then taken away to the next stop—the polishers. Over 400 of these cans are processed every day. They take cans in a terrible state, beyond repair for most, clean them up and sell them back to the oil companies and direct to local consumers. Companies save a fortune.

"You in the West so easily see the slum as a negative concept," said Sonu Badal, a spokesman for Chirag, a campaign to save Dharavi from demolition. "Yes, it is beset by deep poverty and neglect, but Dharavi has also been mirroring India's economic revival and it has done so largely by recycling its own resources."

He continues: "These slum dwellers use their imagination and work hard to make something out of the day-to-day objects others leave behind; Dharavi is an extraordinary success story, its recycling industry employs over 250,000 people."

According to Friends of the Earth, which criticized Northumberland council for sending its waste to India for recycling, Britons should learn more from the developing world.

"In the UK, we could do worse than looking to the recycling endeavour of slum-dwellers in places like Dharavi for inspiration. The level of recycling in the UK is deeply disappointing," says Claire Wilton, of Friends of the Earth. "There is a lot to learn from the developing world, where a scavenger mentality, grass roots recycling and sheer necessity can lead to imaginative leaps in redeploying waste."

In the so-called '13th compound' Dharavi's recycling miracle is in full show. This is where 80 per cent of Mumbai's plastic waste is given a new life. All around young boys cart wheelbarrows filled with everyday plastic waste. Junk is a word that does not exist. Dharavi's plastic recycling industry employs almost 10,000 people, melting, reshaping, and moulding discarded plastic. Close by you will find, the soap-makers, who reprocess soap from hotels and schools. In single rooms hundreds of men toil in the heat over large metal troughs filled with sinister-looking yellow-green liquid. Around them, their co-workers boil vats of molten soap, stirring the cauldrons with oar-sized sticks.

> **Man has lost the capacity to foresee and to forestall. He will end by destroying the earth.**
>
> ~Albert Schweitzer, quoted in James Brabazon, Albert Schweitzer

Dharavi remains a land of recycling opportunity for many rural Indians. The average household in Dharavi now earns between 3,000 and 15,000

rupees a month (£40£200), well above the agricultural wage levels. The new money through recycling has in effect spawned new slum gentry. Certain corners of Dharavi have even gone upmarket with bars, beauty parlours, and clothing boutiques. A major bank has also opened the first ATM in the slum.

But the future of the slum is uncertain. The government has provisionally approved a plan called 'Vision Mumbai' to create a world-class city by 2013. But internationally renowned architect Charles Correa, who has worked in the city for 50 years, says: "There's very little vision with this plan. They're more like hallucinations." Demolition work has begun and police are forcing out inhabitants, leaving thousands homeless. Author and architect Neera Adarkar is among hundreds of activists who see Vision Mumbai as impractical and inhumane because it ignores both the industry and hope of the slum. (Dan McDougall, *The Observer*, 4 March 2007)

Chandigarh's 'Rock Garden': A tribute to recycling

Just in case you forgot, Nek Chand's famous 'Rock Garden' in Chandigarh is a glorious tribute to art, human ingenuity, and the value of recycling. It is an outstanding example how beauty can be created from unexpected materials, like trash and industrial waste.

Started in the late 1950s, Nek Chand began the Garden by using anything he found lying around including stones. He used these trashy materials to sculpt human and animal figures. More than 5,000 pieces depicting birds, monkeys, tigers, soldiers made from trash greet the visitors every day.

Today, the Rock Garden is all of 25 acres. It has works made up of old tube lights, rusty oil drums, broken tiles and china, sanitary ware, glass bangles, and building materials. Nek Chand has even used broken street lights, bricks electrical fittings and wires, soda bottle caps, and even human hair from barber shops.

Recycled plastic roads

K Ahmed Khan came up with an idea to mix recycled plastic with asphalt to lay roadways. Khan has been running his company, KK Polyflex, for 20 years, producing plastic sacks.

About eight years ago, he realized that the anti-plastics lobby had a point, and that the industry was ignoring the problem of plastic waste. He rolled up his sleeves to create an opportunity out of the situation. The plastic appears to have strengthened roads by enhancing asphalt's bonding ability, and made the roads longer lasting by rendering them more resistant to water.

"In 15 days we collected 18 tonnes of waste," says Ahmed Khan. Barring rigid plastics, they accept all manner of film waste without the need for

sorting. He has a contract to lay 500 miles of roads in Bengaluru, where the suitability of the new roadway will be closely monitored. (Good News India)

~John Drinkwater

It's not enough

Before we go and get all misty-eyed with the glories of these success stories there is one important fact to remember. Despite the efforts of various people and organisations, not enough is being done to eradicate waste and its attendant problems.

We must keep in mind that most waste recycling efforts are carried out by private and non-governmental organisations. Therefore, the good and positive effects are very limited and local in nature.

The potential for recycling waste in India is really huge. The prospective opportunities could benefit the entire country if only the country could organise an effective, efficient recycling system, and galvanize people to take it up seriously.

Even as we recycle more, we must remember that it is essential to reduce waste and try and recycle better. Also, it is up to each and every one of us.

SNACKS
CHIPS

8

You can do it too

Recycling is a good thing

When you recycle or compost, you keep stuff out of the landfills and give them a new lease of life. It is relatively easy to start up and sustain a recycling programme, as we aim to show you in the following pages.

In the UK, the proportion of total waste sent to landfills annually has decreased by nearly a quarter in recent years, but in that time *textile waste* has risen to more than one million tonnes, driven by the tendency to discard low cost clothes quickly.

Get started

Whether at work, school or home, you could save a lot of money from recycling. A small effort from your end could make a big difference in the lives of your colleagues, your neighbours in the community, and your family.

When you start a recycling programme, first, find out what you can recycle. It could be everyday items like glass, paper, plastics or leaves, and clipped grass from the garden.

At home

Baby steps

You cannot save the entire planet in one day. So take baby steps, start out small, and do the simple things. You learn a lot too as you take the journey. It is better to start small and do it right the first time.

Recycling at home does not take too much time or effort. Yet the results and benefits can be enormous for you, personally, and for the community at large.

Find out what is recyclable. If you are not sure what to recycle, make a list. There is plenty of helpful information on the Internet that will tell you what you can recycle.

It would be best if you started with one item, for instance paper. As time goes by and you learn more and get better at it, you could add other items, one-by-one to your list.

> **If only...**
> ... the US increased its recycling rates to 60 per cent; the equivalent of 315 million barrels of oil per year could be saved.

Make a note of what most often goes into the garbage in your home. Then sort and separate them into different bins. Two or three bins are generally enough. Remember that sorting and separating are very important in the recycling scheme of things.

You could think of your rubbish bins as your recycling centre. Place them at the back of the garden or in the garage. Find out from your local municipal corporation what they can pick up and what you could deliver to their recycling plant.

Some home recycling information and background
The following tips on recycling could help you on your way to being a good earth citizen.

Paper
Newspapers should be stacked or stored separately from other paper forms because they go straight back to newspaper recycling plants.

Magazines, glossy newspaper inserts, envelopes, and wrapping paper should be separated from newsprint, as they require different recycling methods. Even plastic-lined packs are recyclable.

Cardboard is a highly prized recycling item. These should be stacked separately. Remember to keep all the different forms of paper dry.

Did you know that ...

Recycling a four-foot stack of newspapers saves the equivalent of one 40-foot fir tree.

Plastic

Plastic is an amazing material. It has changed our lives in myriad ways. And yet it is also the cause of most of the garbage and recycling problems of the planet. It also makes up the largest component in landfills. Plastic, in all its forms and varieties, does not breakdown naturally. It just sits there in the garbage dump, floats on rivers, streams, lakes and ponds and lines the sides of roads all across the country.

Plastic, however, can be recycled indefinitely. It can be turned into many, many different and varied products. Call up or visit your local recycling centre (or agency) and find out, what kind of recycling waste they handle. Another good practice to follow is to buy only plastic items

that can be recycled. On the other hand, you should avoid products that use non-recyclable plastic containers.

Plastic by the numbers

There are so many different kinds of plastics that the industry has taken to giving it numbers just so that they (and you and I) know what grade each one belongs to, when recycling, manufacturing, or developing.

To give you an idea here is a list of the different kinds of plastics and their codes and, importantly, their general uses.

- **#1 (PET) and #2 (HDPE) for containers**
  This variety is the most sought after and recyclable in the plastic family. All plastic recycling plants will accept this type. The most common usage for this type of plastic is bottles. This grade of plastic is valuable because it can be spun into a sort of fabric, called 'fleece fabric'.

  The fabric that is made from #1 PET plastic is warm, weather resistant, light-weight, and strong. Did you know that the jacket you bought last winter was made from recycled soft drink bottles?

- **#2 (HDPE)**
  The bags you get at the supermarket and grocery store are generally made from polyethylene—a recyclable plastic.

- **#4 (LDPE) for bags**

- **#5 Polypropylene**
  This is one of the least recyclable of the plastics family. It is generally used to package cheese, vitamins, food wraps, and bottle tops. It is also the cheapest plastic to manufacture and has little or no recycling market value.

- **#6 Polystyrene**
  Polystyrene is used to manufacture cups, food trays, and egg trays. It does not biodegrade and many recycling centres do not take them. So, check with your local recycling plant before sending it to them. Even smarter would be to stop using (or cut down) this kind of plastic product.

- **#7 Polycarbonates**
  This variety of plastic is not recyclable. It is widely used in the manufacture of baby drinking bottles. Studies have shown it is linked to cancer.

Some quick recycling bytes

- Paper takes up as much as 50 per cent of all landfill space.
- Up to 90 per cent of recycled glass can be reused to make new glass items, such as bottles and jars.
- Every glass bottle recycled saves enough energy for a 100 watt light bulb to be lit for 4 hours.
- Americans throw away enough aluminium to rebuild the entire commercial airline fleet every six months.
- Recycling one aluminium can saves enough energy to run a TV for 3 hours.
- Around 36 recycled plastic bottles can make one square yard of carpet.
- About 130 billion beverage containers are sent to US landfills each year.
- Recycling a one-gallon plastic milk jug will save enough energy to keep a 100-watt bulb burning for 11 hours.

Whenever it is possible, try to buy products that are packaged in plastic #2 (HDPE). It is a better and more expensive type of plastic but it is far more acceptable than the other varieties.

Glass

Glass is a highly recyclable item. All glass can be recycled and is used in myriad ways. There is, however, an important thing to remember. Glass needs to be sorted according to colour. This is because it is recycled according to colour. The different colours can be categorized as clear, green, and brown. This class of glass is called 'container glass.'

Glass varieties like light-bulbs, mirrors, pyrex, and sheet glass should be stored apart from the other glasses, such as bottles. This is because in

the recycling process different glasses have different melting points. They also have different elements in their make-up.

> According to a report by the Ministry of Environment and Forests, the plastics industry is growing at 10% per annum, and almost 52% of this is expected to be used in the packaging sector.

Cans

Just like glass and plastic, there are different varieties of cans. They have different compositions and are used to package different products. Food cans are recyclable though they are difficult to deal with.

The most recyclable variety of cans is those that are made of aluminium. However, paint cans and aerosol cans have to be handled carefully. They are often toxic and sometimes dangerous. They should be sorted and stored separately—even from other metals.

Even the lowly aluminium foil packaging is a very useful and recyclable item. It can be turned into engine parts and other mechanical items.

For every hundred residents of Delhi, there is one person engaged in recycling.

~Bharati Chaturvedi, Specialist in Organizing Rag-Pickers in CHINTAN, India

Copper

Copper is probably one of the most reusable and recyclable products. It is 100 per cent recyclable and infinitely so. The best part about recycling copper is that it only requires 15 per cent of the energy needed to produce brand new copper.

Electronic items

Computers, printers, and other peripheral articles may be donated to a local school. Many NGOs work with poor, disadvantaged children or students with disabilities and in the rural areas and find a use for these items, even if they have to cannibalize them.

A subsite on eBay called Rethink Initiative helps people, and businesses, to figure out how to dispose off electronic items, recycle them, and even mend them. The site also explains how to make use of computers that have become obsolete.

Recycling at the office

Start small

Recycling efforts could be started in offices as well. It is best to start out small and not try to do everything all at once. Do what is easy. Firstly, get the staff familiar with the concept of recycling. Even easy things require commitment in making changes in people's daily habits, and habits die hard.

One very important aspect to keep in mind when introducing a recycling programme in the office is that everyone should be involved. There should be a clear communication, setting out the programme in simple terms. Also, make sure everyone understands.

Keep it simple and cheap

First, do a little research. You should organize and build your office-recycling programme so that it is easy. Another important and key point is to keep the recycling programme as small as possible and manageable. If fellow employees feel that they are overloaded, the programme will never succeed. In this way, you will get your fellow office workers to participate and take ownership as well.

Let us permit nature to take her own way; she better understands her own affairs than we do.

~ Michel de Montaigne

Keep any costs involved low. It will then have a greater chance of getting support from the higher management. They would not like to see productive time being wasted on non-business activity.

If you are going to recycle paper and plastics, the biggest waste components in offices, then set the collection boxes or bins around the office. Place enough of them so that people do not have to carry stuff far. Label the bins accordingly.

Involve others

Build a team that involves people from all departments of the office. They are vital in communicating your recycling objectives. They will also be very important in sorting out any issues that arise.

How to recycle in office

Envelopes: Reuse envelopes when you are sending documents within the office or between departments or different branches of the company. You could have a short message on the envelope that explains that the envelope is being reused to save trees.

Paper: Get every one to use both sides of the paper when printing or reuse the paper by cutting them to a smaller size and using them as scratch pads.

Toner Cartridges: Buy only recycled toner cartridges. Send the spent toner cartridges to be recycled.

Coffee Mugs: Buy a whole lot of mugs for use in the office. Avoid use of throwaway or disposable coffee cups, this simple effort would make a big saving for the company and the environment.

Recycling Awards: Institute an awards programme. This is a great way to get people to recycle. Make the awards a public affair because recognition is an important way to get a person to participate, and also make them feel they are doing something good, which is worth appreciation.

Get creative: While recycling is an end in itself, you could get creative with the project. For instance, earnings from recycling were used to set up an employees fitness centre at Intel. Similar steps could be taken by an organization.

> **The earth we abuse and the living things we kill will, in the end, take their revenge; for in exploiting their presence we are diminishing our future.**
>
> ~Marya Mannes

The Möbius Loop

The Möbius Loop is now 30 years old. Gary Dean Anderson created it for the first Earth Day in 1970. Anderson was a young student of the University of Southern California, Los Angeles.

The story of the now universally recognized symbol for recycling is an interesting one.

The Container Corporation of America (CCA) is a paperboard company. They were very aware of the growing impact that the environmental movement was having on people. The CCA was already producing their products from recycled paper. They were one of the first companies to realize the value of recycling and how it minimized the impact on the environment and natural resources.

So, they came up with a two-pronged strategy. One was to raise awareness and promote conservation practices. The other was to be seen as a conscientious and responsible business. The company sponsored a countrywide art contest for a design that would both identify their products and at the same time link it with being a company that used recycled products in their manufacturing. The winning symbol would represent the process of recycling paper. Gary won the competition and a prize of $2500.

The design that Anderson came up with was based heavily on the Möbius strip. The Möbius strip can be described simply as a continuous loop having only one side and one edge. It is both finite and infinite,

simultaneously. Anderson created his design completely by hand before the advent of computers and design graphics.

The Möbius Strip

The **Möbius strip** or **Möbius band** is a surface with only one side and only one boundary component. The Möbius strip has the mathematical property of being non-orientable. It can be realized as a ruled surface. It was discovered independently by the German mathematicians, August Ferdinand Möbius and Johann Benedict Listing in 1858.

The shape of the Möbius Strip probably dates to ancient times. An Alexandrian manuscript of early Alchemical diagrams contains an illustration with the visual proportions of the Möbius Strip. This image, on a page titled 'The Chrysopoeia of Cleopatra', has the appearance of an Ouroboros, and is referred to as the 'One, All'.

[Source: Wikipedia]

 **When we heal the earth, we heal ourselves.** 

~David Orr

> ## What a waste!
>
> According to a survey by the IRG Systems South Asia, a subsidiary of IRG, Washington DC, USA, the total waste from electronic and electrical equipment in India has been estimated to be 146,180 tonnes per year.
>
> Mumbai at present tops the list, followed by Delhi at 9730 tonnes, Bengaluru 4,648 tonnes, Chennai 4132 tonnes, Kolkata 4,025 tonnes. Ahmedabad 3,287 tonnes, Hyderabad 2,833 tonnes, Pune 2,584 tonnes and Surat 1,836 tonnes also figure in the list.
>
> *[Source: IRG Systems, South Asia]*

Oh! That magic feeling

Recycling should be considered as an environmental activity that is responsibility of every one of us, whether at home or at work. Our individual efforts can significantly reduce the negative impact we have on the environment.

So make yourself feel good by doing good. Reduce! Reuse and Recycle!

 Indicates that an object is *capable* of being recycled—not that the object has been recycled.

 Möbius Loop with percentage shows the percentage of recycled material contained in the product.

Where to go and who to ask

Here are some important organizations that one may contact if one wishes to learn more about recycling or even if one wants to get involved.

Centre for Science and Environment

41, Tughlakabad Institutional Area,
New Delhi - 110062
Tel: +91-11 29955124/125; 29956394
Fax: +91-11 29955870; 29955879
E-mail: cse@cseindia.org

State Pollution Control Boards

http://www.indiaenvironmentportal.org.in/directory

India Environmental Portal

This is a very good website that gives plenty of information on recycling and contains a directory of organizations that are involved with all forms of recycling.

Eparisara

A non-government initiative situated in Bengaluru, Karnataka.

Centre for Sustainable Development (CSD) from Bangalore

Nokia Care Centre

To recycle your phone battery or charger all you have to do is drop it off at any Nokia Care Centre. For details, visit http://www.nokia.co.in/get-support-and-software/repair-and-recycle/care-center-locator

Notes
(for further reference)

Chapter 1:
A load of rubbish — what a waste!

www.swa.org/pdf/history_of_garbage.pdf
http://www.swa.org/pdf/activity_sheets/recinnature.pdf
http://www.euwfd.com/html/hydrological_cycle.html
http://inventors.about.com/od/pstartinventions/a/plastics.htm
http://factoidz.com/why-is-anybody-using-trash-bags/
http://inventors.about.com/od/pstartinventions/a/styrofoam.htm
http://factoidz.com/why-is-anybody-using-trash-bags/

Chapter 2:
Just got to have that!

http://www.economist.com/node/13354332
http://www.uow.edu.au/~sharonb/columns/engcol8.html
http://www.grida.no/publications/vg/waste/page/2854.aspx
http://www.cleanair.org/Waste/wasteFacts.html
http://www.grida.no/publications/vg/waste/

http://www.eawag.ch/forschung/sandec/publikationen/swm/dl/global
_waste_challenge.pdf
http://www.indiatogether.org/environment/articles/wastefact.htm
http://www.ceeonline.org/greenGuide/food/upload/environmenthealt
h.aspx

CHAPTER 3:
Reduce—A DIET to save the world

http://en.wikipedia.org/wiki/Jute
http://people.hofstra.edu/geotrans/eng/ch8en/conc8en/ch8c1en.html
http://www.greendustries.com/unido.pdf
http://www.landcareresearch.co.nz/research/sustainablesoc/business/
trade/documents/food_miles.pdf
http://www.vtpi.org/tdm/tdm16.htm
http://www.energybulletin.net/node/5045

CHAPTER 4:
Reuse—Pass it forward

http://www.care2.com/greenliving/why-reuse-beats-recycling.html
http://www.re-nest.com/re-nest/creative-reuse/50-creative-reuse-
ideas-to-keep-you-busy-this-weekend-122566
http://www.recycling-revolution.com/reuse-trash-ideas.html
http://kids.niehs.nih.gov/explore/reduce/reuse.htm
http://www.kidzone.ws/plans/view.asp?i=150
http://www.veengle.com/s/waste%20reuse.html
http://india.alibaba.com/country/products_india-resuable-bags.html
http://www.squidoo.com/reuse-everything

Choose to Reuse, by Nikki & David Goldbeck. Copyright (c) 1995 by Nikki
& David Goldbeck; Ceres Press.

CHAPTER 5:
Recycle — it's the circle of l ife

http://www.history.com/topics/earth-day
http://www.cwc.org/gl_bp/gbp4-0501.htm
http://www.concreteideas.com/recycled-glass-in-concrete
http://biophile.co.za/gardening/start ing-a-vegetable-garden
http://eatdrinkbetter.com/2009/04/03/compost-101-dont-start-a-garden-without-it/

CHAPTER 7:
Recycling is a success story

http://www.epa.gov/osw/conserve/tools/localgov/success/index.htm
http://library.thinkquest.org/11353/gather/stories.htm
http://garbageismoney.wordpress.com/recycle-reuse-respect/
http://www.indiaenvironmentportal.org.in/news/mineral-foundat ion-init iates-solid-waste-management-virdi-smc
http://condofeed.com/2011/12/12/indian-condo-recycles-organic-waste-grow-300kg-grapes/